'Already renowned as pioneers in the history
women, the Rayner-Canhams have now produc
account of chemistry teaching in girls' schools I
nineteenth and twentieth centuries. Based on ε
archival research, this much-needed and informative analysis also offers
a treasure-trove of lively, original quotations from students, teachers and
school magazines.'

Dr Patricia Fara, Clare College, Cambridge

'The authors' painstaking, but elegant, study of the archives of some
sixty independent British girls' schools reveals a forgotten world of
dedicated women science teachers, enthusiastic pupils and well-equipped
laboratories. With its profiles of large numbers of female teachers whose
dedicated service made scientific careers for women possible, *A Chemical
Passion* is a valuable work of reference as well as an absorbing narrative
with captivating illustrations.'

**William H. Brock, Emeritus Professor of History of Science,
University of Leicester**

'An innovative and important book. Drawing upon a range of resources
the authors identify some early women teachers of chemistry and reveal
the excitement and enthusiasm with which girls responded to their
pioneering encounters with the subject. It adds a new and personal
dimension to our understanding of the history of girls'
scientific education.'

**Edgar W. Jenkins, Emeritus Professor of Science Education Policy,
University of Leeds**

A Chemical Passion

IOE Press

A Chemical Passion:

The forgotten story of chemistry at British independent girls' schools, 1820s–1930s

Marelene Rayner-Canham and
Geoff Rayner-Canham

UCL Institute of Education Press

First published in 2017 by UCL Institute of Education Press, University College London, 20 Bedford Way, London WC1H 0AL

www.ucl-ioe-press.com

British Library Cataloguing in Publication Data:
A catalogue record for this publication is available from the British Library

ISBNs
978-1-78277-188-3 (paperback)
978-1-78277-192-0 (PDF eBook)
978-1-78277-193-7 (ePub eBook)
978-1-78277-194-4 (Kindle eBook)

Typeset by Quadrant Infotech (India) Pvt Ltd
Printed by CPI Group (UK) Ltd, Croydon, CR0 4YY

Cover image: Notting Hill High School chemistry room, undated.

Contents

This book has appendices of related material that can be downloaded for free online at www.ucl-ioe-press.com/books/history-of-education/a-chemical-passion/. Their contents is as follows:

Appendix 1: Some chemistry books used at independent girls' schools, 1880s–1920s

Appendix 2: Lives of some women chemistry teachers

Appendix 3: Lives of some pioneering women chemistry students

Appendix 4: The Girls' Public Day School Company chemistry syllabus, 1896

Appendix 5: The Girls' Public Day School Company chemistry syllabus, 1902

Appendix 6: Sophie Bryant's courses of chemistry for girls, 1911

Appendix 7: Rose Stern's courses of chemistry for girls, 1921

Appendix 8: AWST proposed course of chemistry for girls, 1932

List of figures

About the authors

Marelene Rayner-Canham has a B.Sc. from the University of Waterloo, Ontario, while *Geoff Rayner-Canham* has a B.Sc. and Ph.D. from Imperial College, London. They began teaching at Grenfell Campus of Memorial University, in Corner Brook, Newfoundland, in 1975. Marelene retired in 2003 and has since devoted most of her time to research on the history of women in science. Geoff Rayner-Canham is currently Professor of Chemistry at the Grenfell Campus, where his other speciality is inorganic chemistry; with Tina Overton he is co-author of *Descriptive Inorganic Chemistry*, currently in its sixth edition. Geoff has received several awards for excellence in teaching, and he has also taken his chemistry outreach programme to schools in remote and northern Canadian communities.

As well as their major books detailed in the preface, the Rayner-Canhams have also co-authored many research articles on aspects of women in science and have given invited presentations at such venues as the Berkshire conference on the History of Women; the International Union of Pure and Applied Chemistry congress; the biennial conference of the Canadian Coalition of Women in Engineering, Science, Trades and Technology; the US National Women's Studies Association meeting; the American Chemical Society national conference; the American Physical Society annual conference; and the Canadian Society for Chemistry conference.

Preface and Acknowledgements

The genesis of this book can be found in the line of research we have pursued through previous projects. Our first venture into the history of women in science was the accidental discovery of long-forgotten Canadian physicist Harriet Brooks, who had worked with Ernest Rutherford, J.J. Thomson, and Marie Curie (*Harriet Brooks: Pioneer nuclear scientist* was published by McGill-Queen's University Press in 1994). Finding many other women working in radioactivity in the early years, we compiled biographical accounts of 23 women in the field (*A Devotion to Their Science: Pioneer women of radioactivity* appeared from McGill-Queen's in 1997). We were then asked by the American Chemical Society to write a book surveying the history of women in chemistry, the result of which was *Women in Chemistry: Their changing roles from alchemical times to the mid-twentieth century*, published by the Chemical Heritage Foundation in 2001. During our research for that project we identified a very large number of British women chemists active in the first third of the twentieth century, the focus of our subsequent book *Chemistry was their Life: Pioneer British women chemists, 1880–1949* (Imperial College Press, 2008).

The historical question arising from that book was why had there been so many British women chemists in this period? Noting that nearly all of them had attended independent girls' schools led to the research that provides the original source material for this book. We visited four independent girls' schools that some of these chemists had attended – Cheltenham Ladies' College; Croydon High School; King Edward VI High School for Girls, Birmingham; and North London Collegiate School.

During our visits, we became aware that each school had its own magazine, often started in the late nineteenth century. Up until the 1930s, the magazines were academic and chronicled the happenings of the school. They showed that chemistry was being taught even from the early days of the school – in fact, there was more interest in chemistry in those early decades than in the mid-part of the twentieth century. We realized that these magazines provided a unique and hitherto unexplored window into the chemical world of girls' schools. Also, there were staff records and other documents in the respective school archives to provide additional information.

However, the findings for four independent girls' schools could not be taken as an indication of the state of chemistry teaching in that early time frame across the whole of Britain. In fact, there seemed to be a void of knowledge about the early history of the teaching of chemistry in girls' schools. As a result, we were fortunate in receiving funding over several years from the Vice-President (Research, Memorial University)/Social Sciences and Humanities Research Council of Canada Grant, together with periodic grants from the Office of the Vice-President, Grenfell Campus, Memorial University. These grants enabled us to visit Britain for lengthy periods over many years specifically to visit the archives of as many independent girls' schools as possible.

In the autumn of 2011 we were grateful recipients from Dr Sara Delamont of a copy of the monograph *Histories of Girls' Schools* (Barr, 1984). The independent girls' schools for which written histories existed (as of 1984) seemed a good starting point for our choice of schools to visit. As our expertise improved, the large number and diversity of independent girls' schools became apparent. Thus if we were to truly attempt to provide a comprehensive study, we realized that it would be necessary to visit a significant number of schools. Most – but not all – of the schools contacted were very supportive of our research and welcomed us to their premises. In total, we visited the archives of 47 schools and obtained archival information for an additional 15. During the writing phase, our difficulty was not in finding enough material, but picking and choosing from the wealth of information so that we could construct coherent narratives in each chapter.

The complete list of the schools visited is given below. We extend our thanks to the school heads and to the archivists and librarians at each of the schools, without whom this project could never have come to fruition.

Abbey School, Reading
Alice Ottley School, Worcester
Bedford High School
Belvedere Academy, Liverpool
Berkhamsted School for Girls
Blackheath High School
Brighton and Hove High School
Central Newcastle High School, Newcastle-upon-Tyne
Cheltenham Ladies' College
Colston's Girls' School, Bristol
Croydon High School

Downe House School, Newbury
Durham High School
Edgbaston High School
Francis Holland School, Sloane Square
Godolphin School, Salisbury
Hitchin High School
Howell's School, Llandaff
Ipswich High School
King Edward VI High School for Girls, Birmingham
Malvern St James School, Great Malvern
Mary Erskine School, Edinburgh

Maynard School, Exeter

Newcastle-under-Lyme School

North London Collegiate School

Norwich High School

Nottingham Girls' High School

Notting Hill and Ealing High School

Oxford High School

Perse School, Cambridge

Polam Hall School, Darlington

Portsmouth High School

Princess Helena College, Hitchin

Redland High School, Bristol

Roedean School, Brighton

St George's School, Edinburgh

St Leonards School, St Andrews

St Paul's Girls' School, Hammersmith

St Martin-in-the-Fields High School,
     Tulse Hill

St Swithun's School, Winchester

Sheffield High School

South Hampstead High School

Streatham and Clapham High School

Sutton High School

Walthamstow School for Girls

Wimbledon High School

Wycombe Abbey School

There were six other schools that provided us with information but without our visiting.

Bath (Royal) High School

Central Foundation Girls' School, London

Haberdashers' Aske's School, Acton

Manchester High School for Girls

The Mount School, York

Shrewsbury High School

For two more schools, the records were held in archives outside the school and we visited those archives.

City of London School for Girls (London Metropolitan Archives)

St Martin-in-the-Fields High School – early archives (Minet Library, Lambeth)

Eight of the educational institutions no longer existed, though their records survived in archives. The following are the closed schools together with the archive locations we visited.

Cardiff High School for Girls (Glamorgan Archives, Wales)

Esdaile School, Edinburgh (National Library of Scotland)

Leeds Girls' High School (Leeds Public Library, Leeds)

Maria Grey Training College (Brunel University)

Mary Datchelor School (Clothworkers Hall, London)

Milton Mount College (West Sussex Archives, Chichester)

Preface and Acknowledgements

Park School, Glasgow (Mitchell Library Archives, Glasgow, and National Library of Scotland)

Tunbridge Wells High School (Institute of Education Archives, London)

In addition, we thank the Girls' Public Day School Trust for access to the GPDST records held in the GPDSC central offices. We are also grateful for access to the school histories section and the various items in the archives pertaining to girls' schools in the Institute of Education Library. We thank Edgar Jenkins, Emeritus Professor of Science Education at the University of Leeds, for making us aware of the records of the Association of Women Science Teachers at the Brotherton Library, University of Leeds. For the records of the Association of Women Science Teachers, Welsh branch, we appreciated access to the University of Swansea archives. The University of Cambridge main library proved to be an invaluable location for studying school histories, journal articles, and so on. Finally, we thank the library staff at the Grenfell Campus, Memorial University, for tracking down sources and expediting inter-library loan requests.

Lastly, but certainly not least, we thank Nicky Platt, Publishing Director of the Institute of Education Press, for agreeing to take on the publication of our work, and Jonathan Dore, Deputy Publishing Director and Production Editor, for his advice on reworking the manuscript into a much more cohesive and concise narrative.

So, in conclusion, to all the school and archive staff who gave of their time and effort to assist us, we hope that you will find the forgotten early chemistry-life of British independent girls' schools as interesting to read as it has been for us to research, compile, organize, and write.

Marelene Rayner-Canham and Geoff Rayner-Canham

Introduction

Accounts of curricula in English girls' schools in the late nineteenth and early twentieth centuries rarely mention science, particularly chemistry. Science is not addressed in either of Kamm's books: *Hope Deferred: Girls' education in English history* (Kamm, 1965) or *Indicative Past: A hundred years of the Girls' Public Day School Trust* (Kamm, 1971). While in Avery's *The Best Type of Girl: A history of girls' independent schools*, the comment is made: 'In the privately owned schools there was on the whole a marked absence of science before the 1950s' (Avery, 1991: 254). Many other sources can be cited, but what is noticeable is how such statements are presented as 'a known fact' illustrated by: 'Science teaching in girls' schools in the nineteenth century was quite unusual and a laboratory was a rarity' (Sully, 2010: 62).

In her manual, *Discovering Women's History*, Deirdre Beddoe does note the more academic nature of the independent girls' schools, but minimizes the role of the sciences:

> It is worth noting that the education offered in girls' 'public day schools' (i.e. private secondary schools) was far more academic and less sex differentiated than that offered in State schools. Domestic subjects were looked down upon but, on the other hand, science and mathematics were not given the provision which they had in boys' schools in the same sector. Consequently, when many middle-class girls went to university, they had already opted for arts subjects.
>
> (Beddoe, 1983: 59)

An exception to this viewpoint was given by Catherine Manthorpe in a chapter in Geoffrey Walford's *Private Schooling of Girls: Past and present*. Though her focus was on the socio-historical perspective rather than on science, in the concluding points, Manthorpe notes:

> The North London Collegiate School for Girls acted as a model for many of those new schools. From the beginning, science education was included in the curriculum of these schools, and most often one or more branches of the physical sciences were taught as well as mathematics.
>
> (Manthorpe, 1993: 78)

The widely accepted view of a lack of grade-school science education for girls was also at variance with the backgrounds and life stories of early-twentieth-century British women chemists in our own previous research (Rayner-Canham and Rayner-Canham, 2008).

This book, then, is the culmination of our subsequent research into the teaching of chemistry at British independent girls' schools in the nineteenth and early twentieth centuries. Our focus on the independent girls' schools does not imply that chemistry was neglected at all state girls' schools. Though Delamont noted: '... in the state sector things were certainly not so good [for teaching chemistry] ...' (Delamont, 1989: 120). There were some exceptions. For example, in a compilation of girls' school songs, there are photographs of the well-equipped chemistry laboratories at Fulham County School, taken in 1910 (Haddon, 1977: 45) and at Putney Secondary School, taken in 1918 (Delamont, 1989: 120).

There were also co-educational schools at which girls excelled in chemistry, in particular the short-lived (1876–1902) English Higher Grade Schools (Vlaeminke, 2000). For example, in her study of the scientific activities of the suffragist Lydia Becker, Joan Parker noted: '... in 1881 two girls gained first class honours in practical chemistry in the science examinations in the [Manchester] Central Higher Grade Schools' (Parker, 2001: 643).

In this work, to provide a coherent narrative, we focus on those independent girls' schools where we have detailed evidence of a very strong science dimension, particularly in chemistry. Our research is pioneering in its comprehensiveness. The dearth – up to now – of a thorough study of science education at British girls' schools was lamented by Gary McCulloch. Though his comments date from 1987, the points he raised are still valid at the time of writing this manuscript:

> Very little indeed has been written on science education in the girls' secondary schools of the late nineteenth century ... What might now be helpful is detailed research on the social construction of science for girls in the late nineteenth century ... The development of laboratory facilities in secondary schools has still to be charted in detail ... The teaching of specific subjects like chemistry has not yet been inspected in a way that illuminates the cultural context of the schools and the interactions of teachers and pupils.
>
> (McCulloch, 1987: 10–11)

Ruth Watts echoed the same lack of information in her article in 2003. We cite here part of her conclusion and add our own emboldening of the last sentence:

> The history of science and gender is very much a part of our intellectual past and of the history of education. It is vital to understand this past if we are to understand the present state of affairs concerning women in science and work for a better future ... **Unravelling new ways of reading history is essential to challenging unfair assumptions of the past.**
>
> (Watts, 2003: 199)

It is our belief that using the school magazines as primary sources for information about chemistry in schools and about the girls' attitude to chemistry is indeed a **new way of reading history**. In fact, our findings from school magazines have completely overturned the almost universally held view that chemistry was not for girls. As a result, we hope that our research will substantially contribute to the filling of these gaps in the sociological, historical, and scientific records. Our book covers the following topics:

Chapter 1: The revolution in girls' education, 1850–1910

It is crucial to provide a brief context before embarking upon the main function of this book. First, we address some of the reasons for the formation of academic girls' schools. To do so, the political contexts are introduced: the Taunton Report of 1868; the Endowed Schools Act of 1869; the Devonshire Report of 1875; and the Bryce Report of 1895. In addition, the culture for girls changed over the latter part of the nineteenth century, and this too played a role. Then we survey the founding of a selection of the independent girls' schools.

Chapter 2: The earliest chemistry education for girls

Chemistry first became acceptable for girls as a result of the appearance of Jane Marcet's highly popular *Conversations on Chemistry*. Originally published in 1806, the format of the book was a discourse on chemistry between a governess and two young girls. The earliest mention of chemistry lessons that we could find was from 1820 in a Quaker girls' school, while a more substantial account was found for 1824 at a different Quaker girls' school. The archives of The Mount School, York and Polam Hall, Darlington showed that the teaching of chemistry was widespread at Quaker schools.

Chapter 3: Chemistry and the two role-model girls' schools

There were two pioneering institutions which we discuss at this point in order to provide context: North London Collegiate School for Girls (NLCS) and Cheltenham Ladies' College (CLC). As the reader will discover in the subsequent chapters, it was these two institutions, led by their charismatic early heads, Frances Buss and Dorothea Beale, which provided the role models for chemistry taught at the majority of the academic girls' schools.

Chapter 4: Chemistry as a girls' subject

The chemistry course content taught at the schools was comprehensive and thorough, and externally examined. Detailed syllabi were produced and expected to be followed. The chemistry teachers at many of the independent girls' schools adopted the heuristic teaching methodology of Henry Armstrong, what we would today call 'discovery-based' learning. Also described in this chapter is the early-twentieth-century debate on whether academic chemistry or domestic chemistry was the more appropriate for girls.

Chapter 5: The pioneering chemistry teachers

Though it was the headmistresses who determined the emphasis on chemistry, it was the long-forgotten dedicated science teachers who provided the enthusiasm. We look at education of the women chemistry teachers themselves and the challenges they faced. The biographical accounts of some of these amazing women are given in online Appendix 2. Also in this chapter, we summarize the history of the Association of Women Science Teachers, an organization that provided a social forum for female staff.

Chapter 6: Practical chemistry at girls' schools

The key defining issue for the teaching of chemistry was a chemistry laboratory. We discuss the several reasons why the headmistresses of girls' schools pressed for laboratory facilities and show that amongst the academic schools, a chemistry laboratory became a status symbol of an 'up-to-date' school. Though initially, at some schools basement or attic spaces had to be converted into a basic chemistry laboratory, by the 1890s to 1910s, most schools had properly designed and equipped facilities.

Chapter 7: Chemistry and school science clubs

It is from this point on that we look more closely at chemistry from the students' perspective. In this chapter we focus on the science clubs and societies that often had incredibly large memberships of enthusiastic girls. Chemistry-related activities were a major component of the programmes, often including chemical demonstrations and/or outings to chemical plants, which were usually described as 'thrilling'. Many of the clubs and societies peaked in activity either before the First World War or in the inter-war period.

Chapter 8: In their own words: Chemistry poetry and short stories

Of all the parts of this exploration into chemistry at the girls' schools, the greatest surprise to us was the existence of chemistry-related poetry and prose in the school magazines. This incorporation of chemistry into the realm of the arts clearly indicates the depth of chemical interest on the part of these girls. As the reader goes through this chapter, it becomes apparent how much intellect, time, and effort the students put into these literary pieces. In several cases, we show that the poetic works are parodies of well-known classics.

Chapter 9: Chemistry at some Welsh girls' schools

Wales had an autonomous educational system, with its own equivalent government study, the Aberdare Report. The lack of action on the recommendations pertaining to girls' education led to the formation of the Association for Promoting the Education of Girls in Wales. We then focus on chemistry at three academically-focused Welsh girls' schools. The science (particularly chemistry) teachers in south Wales interacted and socialized through the Welsh branch of the Association of Women Science Teachers.

Chapter 10: Chemistry at some Scottish girls' schools

In Scotland, as we demonstrate, chemistry for girls was a key part of an academic training in Edinburgh and Glasgow even back in the early nineteenth century. Like England and Wales, Scotland had its own education report – the Argyll Report. For Scotland, we were able to obtain information on chemistry at four girls' schools. Though the education system in Scotland was totally independent from that of England and Wales, the influences of the GPDSC model for day schools and the CLC for boarding schools held sway north of the border.

Chapter 11: What will the chemistry students do?

Up until chapter 10, we looked at the context of chemistry in the schools. However, what happened to the chemistry-enthused girls after they left school? In this chapter we illustrate how employment opportunities opened up in the late-nineteenth and early-twentieth centuries. Clearly, of the thousands of girls who had a chemistry-focused education, we cannot hope to provide any comprehensive study in this book. Instead we highlight a few individual career paths in online Appendix 3.

Chapter 12: The 1930s: The end of an era

And so to the end of our story. Why did chemistry decline and become forgotten at many of these schools? With the end of the First World War, the momentum for women's advancement was lost and 'traditional' values were promoted. In the context of science teaching, there was one specific 'villain': the often-overlooked Hadow Report. Using 'known' medical evidence, the report concluded that biology was a much more suitable science for girls. The exciting years of chemistry for girls were over.

Chapter 1

The revolution in girls' education, 1850–1910

> *O lift your natures up:*
> *Embrace our aims: work out your freedom. Girls,*
> *Knowledge is now no more a fountain sealed:*
>
> *...*
>
> *Then we dipt in all*
> *That treats of whatsoever is, the state,*
> *The total chronicles of man, the mind,*
> *The morals, something of the frame, the rock,*
> *The star, the bird, the fish, the shell, the flower,*
> *Electric, chemic laws, and all the rest,*
> *And whatsoever can be taught and known ...*
>
> from *The Princess* (Tennyson, 1870: 86, 88)

Among the many early advocates of girls' academic education was Alfred
Tennyson. He was part of the group whose efforts contributed to the
founding of Queen's College (see chapter 3) (Croudace, 1898: 21–2). In
1847, he penned a lengthy poem entitled *The Princess* from which the above
is a very small part. It is noteworthy that Tennyson, a science enthusiast
(Tietze, 1957), included 'chemic laws' as part of girls' education. The poem
itself had a significant impact. It was recited at the opening ceremony
for Queen's College (Croudace, 1898: 21–2), and the phrase 'Knowledge
is now no more a fountain sealed' became a rallying cry for many of the
independent girls' schools.

This book is about our research into the history of teaching chemistry
in British independent girls' schools. But for the ordinary reader, we feel we
need to supply some of the context. Whole books and many research papers
have been devoted to aspects of girls' education in the later-nineteenth and
early-twentieth centuries (see, for example, Avery, 1991; McDermid, 1989;
Walford, 1993; Zimmern, 1898). Here, in this chapter, we have selected
small parts of the narrative that we believe will aid in the comprehension of
the later chapters.

Some 'ancient' girls' schools

To our knowledge, at least three girls' schools date back to the 1600s. The earliest still surviving is Red Maids' School in Bristol, founded in 1634 (Sampson, 1908; Vanes, 1992) as a bequest of John Whitson, MP, Mayor of Bristol, and member of the Society of Merchant Venturers. The school was founded to provide a home and education for the 'daughters of deceased or decayed [poverty-stricken] freemen and burgesses' (Sampson, 1908: 22). It would seem probable that his wish for the education of girls related to his children being three girls, all of whom predeceased him, as did two of his three wives. The school was associated with a hospital and was more like an orphan home for up to 40 girls than a school until 1797, when 'a master was appointed for the first time in the history of this school. He was to teach the ten senior girls writing and the first four rules of arithmetic' (ibid.: 43).

Then in 1659, Blue Maids' Hospital School was founded in Exeter (Anon., 2008). The leading figure in the funding of the school was John Maynard as administrator of the estate of Elizeus Hele which was to be used for charitable purposes. Maynard had two sons and four daughters, and it is of note that he left his own bequests solely to his granddaughters (not grandsons). In 1877, the school was renamed the Exeter High School for Girls, then in 1911, renamed yet again as The Maynard School.

The third existing girls' school to be founded in the 1600s was St Martin's Middle Class School for Girls which later became known as St Martin-in-the-Fields High School for Girls (Siddall, 1999). The school was founded by the Anglican Parish of St Martin-in-the-Fields in 1699 as a charity. The Parish Council, supported by the Society for Promoting Christian Knowledge (founded in 1698), were considered radical for their notion that there should be a local school for girls as well as boys.

There were a few more girls' schools founded in the 1700s. These included the Godolphin School, Salisbury, which was founded in 1726 (though the first students were not admitted until 1784) by Charles and Elizabeth Godolphin (Douglas and Ash, 1928). The school's original purpose was the education of eight young orphaned gentlewomen of Church of England parents. The girls were to be taught to 'dance, work, read, write, cast accounts and the business of housewifery' (Baker, 2013: 148–9). The charity was initially administered by William Godolphin, a nephew of the founders, as the church declined to do so, stating that the idea of educating women was 'mere foolery and dreaming' (ibid.: 149).

The early 1800s saw the founding of some additional schools, such as Princess Helena College in 1820 (Clarke, 1991), originally named the

Adult Orphan Institution, under the Presidency of Princess Augusta, second daughter of George III. The college's original purpose was to educate the orphan daughters of officers who had served in the Napoleonic Wars, and daughters of Anglican clergy. The expectation was that, upon completing their education, they would be exceptionally employable as governesses. In 1874, Princess Helena, Queen Victoria's third daughter, became president of the college, which was subsequently renamed after her. Initially located in Mornington Crescent, London, the college was in Ealing for many years, moving to its current location of Preston in Hertfordshire in 1935.

The first truly academic-focused schools were founded in the 1850s: North London Collegiate School for Ladies (NLCS) in 1850 and the Ladies' College, Cheltenham (CLC) in 1854 (these two schools will be discussed in depth in chapter 3). Thus when the 'great awakening' began in the 1870s and 1880s, there were very few options for a quality education for daughters.

The government reports and acts

One of the greatest events in opening academic education to girls was the Taunton Commission Report of 1868. It had been preceded by the Newcastle Commission Report of 1861 on the status of education in the elementary schools and by the Clarendon Commission Report of 1864 whose mandate was confined to a study of the nine 'great' public [private] boys' schools in England.

The Taunton Report of 1868

The terms of reference of the report of the Royal Commission known as the Schools Inquiry Commission [The Taunton Report] were wide-ranging, but gave no indication that the all-male commission members were going to decide to take the issue of girls' schools as a key part of their mandate:

> To inquire into the education given in schools not comprised within the scope of … [the Newcastle and Clarendon reports] and also to consider and report what measures, if any, are required for the improvement of such education, having especial regard to all endowments applicable or which can rightly be made applicable thereto.
>
> (Maclure, 2006: 89)

It was Emily Davies who heard that the Commission might not include middle-class girls' schools in its mandate unless specifically asked to do so. Davies had been a vociferous campaigner for education rights for women, her views being expounded in her book of 1866, *The Higher Education of*

Women (Davies, 1866). She wrote not only to the chief commissioner, but also to every board commissioner. As Margaret Forster has commented:

> If she had not been so perceptive and zealous in doing this it is more than probable that the already overloaded commissioners would have failed to have girls' schools inspected and the whole rotten edifice of what was laughably termed 'girls' education' would never have been brought crashing down.
>
> (Forster, 2004: 147)

But inclusion of girls' schools in the mandate was only part of the reason for the significance of this commission for girls' education. The other major factor was the sympathy of the commissioners themselves for the women's cause. A key figure was assistant commissioner James Bryce, who was only in his late twenties when he was appointed to his post. Gary McCulloch has commented on the contributing factors to Bryce's personality:

> One was his family background, which was Presbyterian, independent and dissenting, and imbued with a distaste for snobbery as well as a strong sense of public duty. It was also one that lay outside both the metropolitan and the provincial circles inhabited by the English aristocracy and middle classes, and beyond the reach of the elite public schools and the many other private secondary schools in England. He was thus able to comment with fresh eyes on the English scene ... Indeed, his reports are unusually frank, often endearingly and even alarmingly so.
>
> (McCulloch, 2011: 604)

As one of the assistant commissioners, Bryce was tasked with visiting schools in the west and north-west of Britain and examining their academic standards. He insisted that his mandate included studying the girls' schools in the region, in addition to the boys'. The depth and thoroughness of his investigations into girls' schools made Bryce's report to the commissioners into a unique document in itself (Anon., 1868). He noted how the headmistresses had problems with parents not wishing their daughters to learn more than 'accomplishments' [music, dancing, needlework].

Bryce attacked the existing minimalist education of schools for girls and argued passionately that it was unfair that girls were denied the network of academic schools available to boys. He therefore proposed that it would be 'most desirable to provide in every town large enough to be worthy of a [boys'] grammar school a day school for girls' (ibid.: 836).

And, in terms of subject areas, he singled out the current 'absence [for girls] of any knowledge of science' (ibid.: 838). His report had an additional 20 pages devoted to *Answers to Questions Relating to the Education of Girls*. One of the letters he received from correspondents proposed: '… endowed schools for girls, or schools in which there are scholarships tenable under certain conditions, are very necessary, and would be greatly appreciated' (ibid.: 887). Many of the subsequent commission findings seem to have had their genesis in Bryce's own recommendations.

The views expressed by the commissioners themselves are so forthright that they must have been extremely sympathetic to the cause of girls' education. They begin with a clear statement of intent: 'We have thought it our duty to inquire separately into the subject of Girls' Schools …' (Anon., 1867–8: 546). The report then enunciated the collective opinion of the commissioners that an educated mother is of greater importance to the family than an educated father. They expressed support for the submission by a Mr Lingen that, even in 1865, a career rather than immediate marriage might be important for many young women (ibid.: 546).

The report reiterated Bryce's viewpoint that parents themselves were the biggest handicap to academic progress for girls:

> We have had much evidence, showing the general indifference of parents to girls' education, both in itself and as compared to that of boys. It leads to a less immediate and tangible pecuniary result; there is a long-established and inveterate prejudice, though it may not often be distinctly expressed, that girls are less capable of mental cultivation, and less in need of it, than boys; that accomplishments, and what is showy and superficially attractive, are what is really essential for them; and in particular, that as regards their relations to the other sex and the probabilities of marriage, more solid attainments are actually disadvantageous rather than the reverse.
>
> (ibid.: 546)

The Endowed Schools Act of 1869

The major result of the Taunton Commission Report was the passage of the Endowed Schools Act in 1869 (Leinster-Mackay, 1987). This act did not address all the shortcomings of girls' education highlighted by the report, but focused on one aspect, the endowed schools. The distribution of the endowed schools did not match the population distribution of the new industrial society – and there were very few endowments for girls' schools.

The act created the Endowed Schools Commission with extensive powers to require endowments of boys' schools to be divided with girls' schools or even reassign endowments from other purposes and direct them towards the support of new girls' schools. The argument was made that endowments were not the property of the dead but the living.

One of the Taunton Inquiry commissioners, Joshua Fitch, is quoted by Sheila Fletcher in *Feminists and Bureaucrats* as saying:

> ... that if nothing else follows from the Schools Inquiry, this may follow: some substantial measure of rectification of what I conceive to be one of the grossest instances of injustice – one of the most unrighteous deprivations, that can be mentioned, that of, it may almost be said, the whole female sex of England, for a very long time past, of any benefit from the ancient educational endowments of the country.
>
> (Fletcher, 1980: 40)

The work of the commissioners continued from 1870 until 1874. These commissioners were not popular in reassigning the 'birthright' of the boys' schools. Fletcher cites a statement by Carey Tyso of Wallingford, who had proposed splitting a boys' school endowment in order to re-establish a girls' school in the community:

> To the Ladies, I may say a parting word. I have exposed myself to adverse criticisms, if not obloquy, for venturing to advocate the claims of the female sex to a share in the Endowment. I accept it cheerfully, feeling confident of your moral support and sympathy, and backed in my opinion by many of the most eminent writers on Education, and distinguished advocates of the progressive enlightenment of all classes of the English Community.
>
> (ibid.: 86)

This issue of reassigning endowments to other schools arose at the hearings of the subsequent Bryce Commission of 1895 (see below):

> Of course, the old expression of Mr Erle's that the Endowed Schools Commission gave power to turn a boys' school in Northumberland into a girls' school in Cornwall is an extravagant statement of the power, and such a change cannot often be made; but the nearest case that I can remember of a change of that kind was when 40,000*l.* belonging to the Aldenham Boys' Grammar School in Hertfordshire was applied to the North London

Collegiate School for Girls in North London. That was literally money taken from boys and applied to girls, and in a different locality ...

(Anon., 1895: 441)

The Endowed Schools Commission was abolished under the Endowed Schools Act of 1874 and its mandate was transferred to the authority of the Charity Commissioners. This body held the power of redistribution of funds until the passage of the Board of Education Act of 1899.

It was the Taunton Commission Report and the ensuing aggressive actions of the Endowed Schools Commission and then of the Charity Commissioners that revolutionized the education possibilities for middle-class girls. Joyce Senders Pedersen commented: 'By 1894, there were over two hundred endowed and proprietary schools for girls, virtually all of them created or recreated in the period after 1850, the majority dating from after 1870' (Pedersen, 1975: 148).

The Devonshire Report of 1875

There had been widespread concern at the lack of science teaching in British schools. As a result, the all-male Royal Commission on Scientific Instruction and the Advancement of Science was formed. The series of reports issued by the commission were named the Devonshire Reports, after the chairman, the Duke of Devonshire. The sixth report of the committee focused upon the secondary schools, or to be more precise, the boys' endowed schools. Their survey showed that of the 128 schools from which they received reports, science was taught in only 63, of which only 13 had a laboratory (Anon., 1875: 4–10).

The report was greeted with outrage, such as that of the editor of *The Chemical News*: 'Hence it would appear that England, in 1875, is about on a level with Germany in the year 1820, before the great impulse given to scientific education by Liebig and his contemporaries' ('Editor, The', 1875: 149). In particular, the report emphasized the need for chemistry and physics: 'We desire to record our opinion that School Laboratories should be constructed so as to supply accommodation for Practical Work in Physics, as well as Chemistry' (Anon., 1875: 5).

Though the report recommendations were to apply to all public and endowed schools, the sole use of the terms 'boys' and 'masters' would imply that the teaching of boys was the only focus. Curiously, the term 'boys and girls' did appear in an earlier related document of 1872. The *Report of the Endowed Schools Commissioners to the Lords of the Committee of Her Majesty's Privy Council on Education* contained a letter to the Endowed

Schools Commission from Lyttelton, Duke of Devonshire, chairman of the Committee on Scientific Instruction (Anon., 1872). This letter laid out the basis for the proposals that subsequently appeared in the Devonshire Report, but here with specific reference to both boys' and girls' schools:

> In all these [endowed] schools without exception, whether for boys or girls, we propose to require, as a substantial and indispensable part of their course of instruction, at least one branch of Physical Science; and in a few intended for the more special encouragement of what may be called *modern* subjects, we suggest, without absolutely requiring, more extensive teaching of science.
>
> (Lyttelton, 1872: 66)

It was the Devonshire Report, then, that essentially required all endowed schools to incorporate at least one physical science (usually chemistry) in their curriculum and to provide the necessary laboratory facilities.

The Bryce Report of 1895

The final commission of relevance in this period was the report of the Royal Commission on Secondary Education, chaired by James Bryce, an obvious choice after his extensive contributions to the Taunton Report. Other commission members included three women: Dr Sophie Bryant (see chapter 3), Lucy Cavendish, and Nora Sidgwick. Lucy Cavendish, née Lyttelton, was president of the Yorkshire Ladies' Council of Education (Fletcher, 2001; Russell and Cohn, 2012). Eleanor (Nora) Mildred Sidgwick, née Balfour, was a physicist and activist for the higher education of women. One of the first students at Newnham College, she was appointed principal of Newnham in 1892 (Fowler, 1996: 7–28).

Illustrating the changes since the Taunton Commission 27 years earlier, and probably indicating the influence of Bryce himself, there were also women assistant commissioners with a special brief to report on girls' education. In addition, there were many women who gave evidence to the commission (Goodman, 1997). The terms of reference of the commission were:

> To consider what are the best methods of establishing a well-organized system of secondary education in England, taking into account existing deficiencies and having regard to such local sources of revenue for endowments or otherwise as are available or may be made available for this purpose and to make recommendations accordingly.
>
> (Maclure, 2006: 140)

There were a few traditionalists among the assistant commissioners, one being a Mrs Kitchener. Kitchener could not comprehend the wishes of many Liverpool girls for practical science classes:

> Mrs Kitchener experienced great difficulty accepting the attraction of this type of science class for girls. She had particular trouble with the desire on the part of the girls in Liverpool for science education in the higher grade school. In Liverpool all the scientific subjects that required apparatus were taught in the Central Higher Grade [boys'] School and the girls were required to walk back and forward for the lessons. As a result, it was thought that the girls would be better having what lessons they could get from their own teachers ... The girls of six schools, however, petitioned the school board asking to be admitted with the boys to the science lessons at the Central Higher Grade School. The board agreed to the girls' request and 120 girls put down their names for the organized science school.
>
> (ibid.: 299)

The independent academic girls' schools

As mentioned above, following publication of the Taunton Commission Report and the implementation of the Endowed Schools Act, a large number of academic girls' schools were established. Also, for those already existing, such as Exeter High School and Princess Helena College, academic subjects were upgraded with the hiring of specialist teachers. In a brief overview of those halcyon years, we provide a brief survey of attitudes to the academic education for girls, as a prelude to our more focused discourse on chemistry in girls' schools in subsequent chapters.

An academic education for girls?

Though parents were often opposed to their daughters acquiring any other knowledge than needlework and music, many girls disagreed. Gillian Avery eloquently described:

> A passionate desire for learning seems to have been a characteristic of many of the early nineteenth-century girls ... With dedicated self-discipline many girls worked at home, teaching themselves Latin, Greek, German, and Italian, reading Dante, translating Schiller, in between the demands that their families made upon them.
>
> (Avery, 1967: 184)

Such a thirst for knowledge on the part of girls continued into the later part of the nineteenth century. Mabell Ogilvy, later Countess of Airlie, whose childhood was spent in the late 1870s, reflected:

> I longed to study science, mathematics, and other subjects barred as 'unfeminine'. The sole approach to my ambition was when my brother Arthur was once given special coaching in Euclid during his school holidays and I was encouraged to share the lessons – with a view to encouraging him. I remember too teaching myself the Greek alphabet in secret, but when this illicit study was discovered the book was confiscated and I was ordered to do some needlework instead.
>
> (Airlie, 1962: 34)

Ogilvy recalled the prejudices among many of the upper-class parents:

> Girls' schools were looked upon with horror in the eighteen-seventies, and none of our contemporaries went to one, with the exception of our cousins, the Ashley girls [daughters of the reformer, Lord Shaftesbury] who attended a High School. They reaped the benefits of a wonderful education, but the fact of their doing so was always deplored, so strong was the prejudice.
>
> (ibid.: 34)

If the parents were often opposed to academic girls' schools, why were the schools so successful? Ellen Jordan has suggested that: 'The parents were prepared to pay for an education for their daughters, but cared very little about its content so long as it guaranteed gentility and femininity' (Jordan, 1991: 450). She added that the educational reformers argued that 'only women educated in the way they favoured could become good wives and mothers and therefore truly feminine' (ibid.: 450). Jordan elaborated:

> This curriculum was intended to create a literate, cultivated woman who might not necessarily have the proficiency in the classical languages provided by the boys' public schools and the universities, but who would probably have a better knowledge of more recent literature, both English and European, and of history, geography, and natural science.
>
> (ibid.: 451)

The role of the school headmistresses

The dedication and fearlessness of the pioneering headmistresses at these schools cannot be overemphasized. And it was often very young shoulders that bore this burden incredibly well, as Nonita Glenday and Mary Price have noted: 'Most of the headmistresses were in their early twenties, very young for the responsibilities they were carrying, and had led sheltered lives' (Glenday and Price, 1974: 23).

These headmistresses (and teachers) met fierce opposition from several directions as Sara Delamont noted:

> The feminist pioneers who opened academic secondary schools for young ladies in the second half of the nineteenth century did so against a body of medical opinion, religious orthodoxy, and a widespread belief among their potential clientele (that is, middle- and upper-middle-class parents) that such institutions were dangerous. A pupil, or a member of staff, at an academic secondary school was held to be in physical danger (her health would suffer, she might become subfertile or die of brain fever); in moral danger (away from the control of her mother she might meet *anybody*) and liable to forfeit her marriage prospects for men who would not want a wife who knew algebra.
>
> (Delamont, 1993: 234–5)

Margaret Bryant called these women 'the agents of change' (Bryant, 1979: 60–75). As Pedersen pointed out, it was these headmistresses who defined the identity of the schools:

> Headmistresses in some cases founded or assisted in the initial organization of their institutions, and it was usually the teachers who initiated curricular and other reforms within the schools. Without the teachers' professional vision to structure the reforms, girls' secondary education might well have developed rather differently than it did in mid- and late-Victorian years.
>
> (Pedersen, 1975: 162)

Academic successes

To be judged as academically successful, it was necessary for the talented girls to be entered into nationally accredited examinations. There were five such bodies: the Oxford and Cambridge Schools Examination Board; the Oxford Local Examinations Delegacy; the London Matriculation; the Science and Art Department; and the College of Preceptors.

Unfortunately, in a review of science examinations around the end of the nineteenth century, the author does not provide separate statistics for boys and girls (Crane, 1959). The information, however, does provide a framework for our own findings. According to Crane, very few students took science examinations with the Oxford and Cambridge Schools Examination Board. Chemistry was taken as a combined paper: 'Heat and Chemistry' (ibid.: 54). But it is notable that in 1904 the board reported that the increase in science candidates was the result of 'the larger number of Girls' Schools examined' (ibid.: 55). Similarly, few students took the chemistry examinations available through the Oxford Local Examinations Delegacy.

For the London Matriculation, science was compulsory. This fact provided some school headmistresses, such as Miss Dove at St Leonards School, St Andrews (see chapter 10), with a powerful weapon to persuade school governors of the need for improved chemistry facilities, particularly laboratories. At many of the schools we visited, the early school magazines listed the external examination results. We noted that, of the sciences taken by girls for the London Matriculation, chemistry was by far the most popular.

Established by the British government to promote education in art, science, technology, and design, the Science and Art Department, South Kensington, functioned from 1853 to 1899 (Wikipedia contributors, 2015). The department offered examinations in all sciences, including separate examinations for inorganic chemistry and organic chemistry. Of all the science examinations, inorganic chemistry had the greatest number of registrants (Crane, 1959: 60–1). From our own research, many girls' schools entered their students in the Science and Art Department examinations. For example, in 1880, Cheltenham Ladies' College had 15 students enter and pass the inorganic chemistry examination – compared with nine students for mathematics (Anon., 1880: 232).

Incorporated by Royal Charter in 1849, the College of Preceptors examined and provided certificates for secondary school students, both boys and girls, in a wide variety of subjects. Crane noted that 'until the end of the [nineteenth] century, there were more candidates for the examinations of the College [of Preceptors] than for the local examinations of Oxford and Cambridge combined … a majority of the schools were for girls …' (Crane, 1959: 58). In these examinations the girls proved to be exceedingly academically capable. Andrea Jacobs has documented the academic successes of girls in the period 1860 to 1902 (Jacobs, 2001). Jacobs reported:

By the end of the decade [the 1870s], the number of girls entering for the Schools' Examinations of the College of Preceptors exceeded boys. In 1879 there were 7,645 candidates of whom 4,428 were female and the following year there were 9,148 candidates of whom 5,121 candidates were female. Not only were there more female candidates but their success rate was often higher.

(Jacobs, 2001: 125)

In fact, as the 1880s progressed, the greater academic successes of girls than boys became an issue of concern. Jacobs commented: 'By February 1884 the gap between the performance of the girls and that of the boys had widened to such an extent that it could no longer be dismissed' (ibid.: 129). However, the conclusion drawn at the time was that the problem lay with the girls: the girls were working too hard. By contrast, it was argued, boys had a healthier, more casual, attitude to their studies. This dedication to, and enthusiasm for, their studies by girls was seen as a cause for concern. For example, it was believed that the excessive use of the brain by girls could have led to lasting damage to their reproductive organs (ibid.: 132). This 'dedication and enthusiasm' certainly applied to girls' attitude to chemistry, as we show in the later chapters.

The New Girl

Not only did society change, but the young women themselves underwent a revolution. The admired attributes of yesteryear of obedience and docility were no more. In 1898, Alice Zimmern wrote: 'The High Schools have produced a new type of girl, self-reliant, courageous, truthful, and eager for work' (Zimmern, 1898: 76–7). There were several factors responsible for this. In particular, Sally Mitchell has contended that the introduction of the school uniform promoted among young girls a liberating sense of 'girlhood':

I wonder, however, whether late-twentieth-century schoolgirls' joy in getting rid of school uniforms when they leave adolescence comes anywhere close to the turn-of-the-century girl's dramatic liberation when she first dressed in a costume distinctly her own, which marked her as neither child nor woman, had pockets, made it possible to run and climb, and let her add a boy-style shirt and tie.

(Mitchell, 1995: 89)

Victorian and Edwardian middle-class girls discovered role models through the new magazines for girls that appeared: *Atalanta* (1887–98), *The Girl's Realm* (1898–1915), and *The Girl's Own Paper* (1880–1927) (Dixon, 1998/9). Also, according to Diana Dixon, school-age girls read of science, including chemistry, in their brother's boys' magazines, or in magazines produced for both boys and girls, such as *Young England* (Dixon, 2001: 235).

Though the viewpoint of the girls' magazine depended upon the editor of the time, these magazines were very forward-thinking and quite adult in their messages. The issues contained a significant proportion of intellectual material, including articles on science topics; reviews of colleges and universities for young women; and the widening career opportunities for women. With the girls' magazines urging them forth, many of the young girls of the 1880s to 1920s saw it almost a duty of their generations to view a university education as their next goal.

The Girl's Realm was the most forward-thinking of all, contending that girls had an entitlement to an education equal to that of their brothers. This viewpoint was exemplified by Alice Corkran's 'Chat with the girl of the period' in the very first volume (Corkran, 1899: 216). In a study of *The Girl's Realm*, Kristine Moruzi explained why it was the most progressive of the girls' magazines:

> In part because it first appeared so late in the nineteenth century – and, therefore, was unlike many other girls' periodicals of the period, which had stronger ties to the Victorian era – the *Girl's Realm* is significantly less constrained by nineteenth-century ideas of femininity. Unconcerned with a nostalgic feminine idea of the past, the magazine instead uses current events to fashion girlhood as a time of bravery and courage. The girls in its pages are educated and feminine and the *Girl's Realm* positions them as heroes. Involved in adventures both at home and abroad, girls are capable and confident when they need to be ... and non-fiction articles encourage girls to seek opportunities beyond the domestic sphere.
>
> (Moruzi, 2009: 241–2).

Some of the content was quite subversive: for example, an issue of *The Girl's Realm* of 1914 carried an article 'The Woman of the Future' which expounded at length that 'One of the greatest social crimes that has ever been committed was man's enslavement of woman.' The author, Orison Swett Marden, anticipated woman's future role:

Woman is taking to herself a new significance. She is discovering that she, as well as man, has another message for humanity besides that presented by wife and mother, and that henceforth she is to fulfil an entirely new mission in civilization. Today woman believes that her life line should run *parallel* with man's. The new woman protests against having her life absorbed in ministering to man, in being exploited wholly for his benefit.

(Marden, 1914: 449)

These girls had their own genre of novels about girls in girls' schools. In the early part of the twentieth century, several series of novels were published that changed the nature of fiction for girls by presenting a young female point of view that was active, aware of current issues, and independently-minded (Reynolds, 1990). Gill Frith has commented:

In a world of girls, to be female is *normal*, and not a *problem*. To be assertive, physically active, daring, ambitious, is not a source of tension. In the absence of boys, girls 'break bounds', have adventures, transgress rules, catch spies.

(Frith, 1985: 121)

The stories were invariably set in upper-middle-class boarding schools, yet with sales of these books far exceeding the number of girls at all secondary schools, they provided the idealistic goal for a much wider audience as to girls' permissible behaviour and ambitions as Eva Löfgren described:

The world of the boarding school story in British girls' fiction in the first half of the twentieth century is an Amazonian community where all the principal roles are played by women and girls. Artemis and Athena are here the ruling deities rather than Aphrodite or Hera. Men are only admitted into this all-female society for menial duties as gardeners and boot boys or as visiting outsiders.

(Löfgren, 1993: 43)

As the reader will discover in later chapters, the quotes of the girls themselves reflect the philosophy of the 'New Girl'. The students become 'enthralled' at doing chemistry; 'excited' by venturing into the chemistry laboratory; and 'thrilled' by going on industrial chemistry expeditions.

The Girls' Public Day School Company Schools

As we have mentioned above, the 1880s to 1900s marked the founding of a large number of independent girls' schools. In particular, a network of day schools was inaugurated that mandated chemistry as part of the curriculum. These were the schools of the Girls' Public Day School Company (GPDSC). Here we provide a brief background.

Few parents could afford the fees – or had the social class – to send their daughters to the upper-class girls' boarding schools. It was to fill this void that members of the National Union for the Improvement of the Education of Women of all Classes (NUIEWC, also known as the Women's Educational Union) organized a public meeting at the Albert Hall in 1872. At the meeting, the Girls' Public Day School Company (GPDSC) – later the Girls' Public Day School Trust Limited – was inaugurated (Goodman, 2004: 26). The purpose of the company was to raise funds for the independent, affordable, academically-selective girls' schools. A GPDSC school was only considered for founding if there was a request from a group of citizens in that particular community.

The organization was run by the GPDSC council. This body governed the network of schools that was formed, while leaving each some degree of independence to grow and develop according to the wishes of the headmistress and of the community. The council did demand that the education supplied to the girls be the best possible and be given at a moderate cost. The aim was clearly enunciated that '... [the education] was to correspond with that provided for boys in the great public schools' (Burstall and Douglas, 1911: 13).

The GPDSC schools were modelled on the North London Collegiate School (NLCS), founded by Frances Buss. In view of its importance to the narrative, a major part of chapter 3 will be devoted to Buss and the NLCS. In fact, in the early years, the heads and assistants of GPDSC schools were required to visit NLCS before taking command of their own school in order to study and reproduce the methods and organization of the NLCS (Reynolds, 1950: 114–15).

The GPDSC schools are as follows, the names being in italic if they subsequently closed, merged, or left the company: Bath (Royal) High School; Birkenhead High School; Blackheath High School; Brighton and Hove High School; Bromley High School; *Carlisle High School*; *Clapham High School*; Croydon High School; *Dover High School*; *Dulwich High School*; *East Liverpool High School*; *Gateshead High School*; *Hackney (and Clapton) High School*; *Highbury and Islington High School*; Ipswich High School;

Kensington High School; Liverpool (Belvedere) High School; *Maida Vale High School*; Newcastle High School; *Newton Abbot High School*; Norwich High School; Nottingham Girls' High School; Notting Hill (and Ealing) High School; Oxford High School; Portsmouth High School; Putney High School; Sheffield High School; Shrewsbury High School; South Hampstead High School; Streatham Hill (and Clapham) High School; Sutton High School; *Swansea High School*; Sydenham High School; *Tunbridge Wells High School*; *Weymouth High School*; Wimbledon High School; and *York High School*.

The NLCS was a model for many other academic day girls' schools, including the Manchester High School for Girls. To illustrate, in 1900, there was a formal three-day jubilee celebration of the founding of NLCS, including a service at St Paul's Cathedral. An account of the several events in *The Magazine of the Manchester High School* ended with:

> Thus was brought to a close one of the most interesting celebrations of our time – the Jubilee of the first 'High School for Girls,' the success of which has meant so much for all of us; for to Miss Buss's efforts we owe the inception of the whole movement, and were it not for her valiant struggles against much opposition, the establishment of Girls' High Schools, and the opening of the Universities to women, might have been greatly retarded.
>
> (Anon., 1900: 64)

Some other academic girls' schools

Though the GPDSC provided a network of academic girls' schools – particularly in the Greater London area – many other organizations set up schools. It is not our purpose to provide a comprehensive listing of the more than 100 academic girls' schools formed in the later 1800s and early 1900s, but to introduce five more of the day girls' schools that were particularly science-active and which feature prominently in our subsequent chapters. In particular, it is of note that each of these schools had some link to Buss and the NLCS.

Manchester High School for Girls

It was the Manchester Association for Promoting the Higher Education of Women that initiated the process leading to the formation of the Manchester High School for Girls (Goodman and Harrop, 2002: 38–9). The association resisted affiliation with the GPDSC and instead set up a provisional committee of ten women and eleven men. The committee undertook fundraising announcing: '... such a School is a pressing need in this city in particular, and would still only scantily provide for Manchester's daughters what has been provided without stint for Manchester's sons' (Burstall, 1911: 4). A building was acquired and a headmistress appointed, Elizabeth Day, while the school opened in 1874.

In 1881, the Governors insisted on systematic science teaching to be an integral part of the school programme (ibid.: 106). However, Day had 'grave doubts as to the suitability of Science teaching, especially Chemistry as then taught ...' (ibid.: 106). It was to be Day's successor as headmistress, Sara Burstall, who placed a strong focus on science.

Burstall was another influential proponent of girls' education. Born in Aberdeen in 1859, she attended Camden School for Girls. The Camden School had been founded by Buss for families who could not afford the fees at NLCS (Burchell, 1971). Burstall obtained a scholarship enabling her to complete her high school education at NLCS, following which she entered Girton College, Cambridge, in 1878. Completing the Tripos in 1882, Burstall returned to NLCS to teach. With the encouragement of Buss, she completed a BA degree at the University of London in 1884. Following the death of Buss, Burstall was appointed deputy headmistress at NLCS in 1894. Burstall then accepted the position as headmistress of Manchester High School for Girls in 1898, retiring in 1924 (Burstall, 1933; Delamont, 2004).

King Edward VI High School for Girls, Birmingham

The King Edward VI High School for Girls (KEVI) was founded as a result of redirecting endowments. The original foundation funds of 1552 came from the seizure of monastic lands and were used to endow a boys' school. It was not until 1883 that the Charity Commissioners (see above) enabled the transfer of funds to form the equivalent girls' school.

The first headmistress at KEVI, Edith Elizabeth Maria Creak (1855–1919), completed the Cambridge Tripos at Newnham, was first appointed as assistant mistress at Plymouth High School in 1875, then as the first headmistress of the GPDSC Brighton and Hove High School in 1876 at age 20, and finally to KEVI in 1883. She was an enthusiastic supporter of science for girls. The earliest KEVI school biographer, Winifred Vardy, noted:

> To Miss Creak belongs the honour of being a pioneer in the teaching of Science to girls. Though her own training [at Newnham College, Cambridge] had been Mathematical and Classical, she seems to have foreseen the value of scientific knowledge for women ...
>
> (Vardy, 1928: 25)

Central Foundation Girls' School

The Central Foundation Girls' School can trace its origins back as far as 1697 to small co-educational charity schools. The key figure in founding the academic Girls' School was the Revd William Rogers (Anon., 1892: 24–5). In 1866 Rogers founded the Middle Class Schools Corporation with the intent of opening a boys' school and a girls' school. Despite the fact that significant moneys had been raised and a building acquired for the girls' school, many of the corporation council members were horrified at the thought of bringing girls in by rail from the suburbs and collecting them in a large public school. They imagined all types of terrible consequences and declined to undertake the responsibility. The scheme was abandoned and the moneys for the girls' school returned to the donors.

By 1891 the climate towards girls' education had changed dramatically and the founding of the Central Foundation Girls' School that year finally brought Rogers's ambition to fruition (France, 1970). The second headmistress, Mary Hanbidge, appointed in 1898, had been educated at Camden, NLCS, and CLC. She was an equally fervent supporter of science for girls and ensured that science, and chemistry in particular, was a core part of the curriculum.

Mary Datchelor School

The 'now-deceased' Mary Datchelor School also owed its origins to the Taunton Report and the Endowed Schools Commission. A large sum of money had accumulated as a result of outdated bequests from the estate of Mary Datchelor and her sisters (Anon., 1957). The commissioners appointed trustees with the implicit instruction to reassign a significant portion of the derelict funds to found a day school for middle-class girls of the parish. The school opened in 1877 and was a great success, enrolment rising to 500, but it was underfinanced until the Clothworkers Guild took over in 1894. Even in such an autonomous school, it is interesting to note that the five images in the stained-glass window in the school hall included Frances Buss of NLCS, together with Maria Grey and Lady Stanley, two of the founders of the GPDSC.

The headmistress, F.E. Grimshaw, was particularly proud of the performances of the students in their chemistry examinations. For example, an issue of the *Datchelor School Magazine* of 1901 proudly reproduced the report of the chemistry section of the high school examination administered by Cambridge University: 'In the Upper VI [class] the section on Physical Chemistry studied had been thoroughly mastered, and a large number of girls scored over 90 per cent of the marks' (Anon., 1901: 16). Some of these students proceeded to university to study chemistry. For example, in 1908, two students from Mary Datchelor School entered the (women's) Royal Holloway College of the University of London, to study chemistry: 'She [Edith Hancock] is entering for an Honours Degree in Science, taking Chemistry as her special subject. Phœbe [Routh] is also reading for Honours in Chemistry ...' ('C.E.R.', 1908: 5).

In the 1970s, pressure was put on the school administrators to merge with a boys' school. It was felt that the merger would result in the end of the unique atmosphere of Mary Datchelor, however, and it was decided to close the school instead. Closure came in 1981.

Milton Mount College

The wide range of chemistry-active schools was astounding. To provide just one other example, we have chosen Milton Mount College (another of the 'deceased' schools). The college was founded in 1871 at Milton, near Gravesend, Kent, for the education of daughters of Congregational ministers. And it was an academic education they planned to give, as stated by the first lady principal, Miss Hadland, at the opening of the college: 'We have very much at heart the service which we can render to ministers' daughters who are looking forward to self support' (Binfield, 1981: 28). In the audience that day was Frances Buss, indicative of the role model that Milton Mount intended to emulate (ibid.: 29).

Figure 1.1: Milton Mount chemistry laboratory, undated
Source: Archives, Milton Mount College

The school biographer, Hilda Harwood, commented: '... in 1877 some groups of the Cambridge Higher Local Examinations were successfully attempted. Botany was now taught and in the following year classes were begun in chemistry, geology, and physiography' (Harwood, 1959: 12–13). A chemistry laboratory was constructed as part of the original building plan, however, it was not until 1883 that the laboratory was fully equipped as a result of 'the school's connection with the South Kensington Science Department ...' (ibid.: 13).

The War Department appropriated the school building for the duration of the First World War. By the end of the war, the structure had been so badly neglected and damaged that it was no longer fit for use as a school. Milton Mount found a new home near Crawley in Sussex, but was finally forced to close in 1960. Clyde Binfield commented:

> Schools for ministers' daughters seemed less relevant in the context of an efficient national system of secondary education ... The academic edge too had gone as opportunities for able Nonconformist girls became as wide as those for any girls: the succession of Milton Mount girls at Newnham was not maintained. What was a frontier post in the 1870s had become a last bastion a hundred on. The school had been too successful.'
>
> (Binfield, 1981: 29)

In this chapter we have provided a brief background to the great expansion in academic girls' schools. It was the Endowed Schools Commissioners and later the Charity Commissioners (deriving their authority from the Taunton Commission report) who unlocked the endowments that led to the founding of many of these schools. And fortunately, there were many women willing to take up the challenge of running them. Having described the context, we now look specifically at the teaching of chemistry at independent girls' schools.

Chapter 2

The earliest chemistry education for girls

There's a dear old place in the suberbs [sic] of York
That goes by the name of the Quaker Girls' School
...
We'll always remember the lab and the gym
The former for odours the latter with pride.

(Anon., 1892: n.pag.)

As we show throughout this book, many rhymes written by students included some chemistry-related content. This particular ditty is appropriate to open this chapter as it comes from one of the Quaker girls' schools discussed in this chapter. First, however, we describe the book that contributed to making chemistry an acceptable subject of study for girls: *Conversations on Chemistry*.

Jane Marcet's *Conversations on Chemistry*

Arguably one of the key reasons for the acceptability of chemistry for girls was the existence of a text written specifically for young women. The book, *Conversations on Chemistry: In which the elements of that science are familiarly explained and illustrated by experiments*, was authored by Jane (Haldimand) Marcet (Marcet, 1817), though the early editions were anonymous.

Books on science for children first appeared in the latter half of the eighteenth century. They were designed to appeal to children's innate curiosity while at the same time reinforcing a belief that such amazing phenomena were the work of the 'Creator of all things' (Fyfe, 2008). A popular format was that of a fictional conversation between a parent or tutor, and two children. By this means, the child-reader could identify with one or other child in the book. In addition, the interplay between the three characters provided a different dimension of interest. Dramatic scenes of discovery added to the readability of such books. It was Maria Edgeworth, assisted by her father, Richard Lovell Edgeworth, who noted in their 1798 book, *Practical Education*, that the conversational style was particularly applicable to the teaching of the sciences (Edgeworth and Edgeworth, 1798: vi).

Marcet's book became the classic (and only) book on chemistry specifically to appeal to teenage girls, and 18 editions appeared in Britain, the first in 1806 and the last in 1853 (Cole, 1988: 420). She contended that the conversational style was 'more especially [appropriate] to the female sex, whose education is seldom calculated to prepare their minds for abstract ideas, or scientific language' (Marcet, 1817: vii).

The conversations themselves consisted of a series of discourses by a Mrs B. and two students: Emily, who is serious and hard-working, and Caroline, who is more spontaneous and imaginative. Caroline's initial lack of enthusiasm is summed up by her first statement in the book: 'To confess the truth, Mrs. B., I am not disposed to form a very favourable idea of chemistry, nor do I expect to derive much entertainment from it' (ibid.: 2). The characters are sketched only superficially, the two students being well-educated town-dwellers between 13 and 15 years of age from wealthy families. Mrs B. is identified as Mrs Bryan. As a prominent scientific author of the time was Margaret Bryan (Alic, 1986: 177), it is possible that the anonymous Marcet wished to lay a false trail as to the authorship of *Conversations on Chemistry*.

Marcet was uniquely qualified to write the book. Jane Haldimand was born in London in 1769, the daughter of affluent Swiss parents (Bahar, 2001). At 15 years of age her mother died and she had to take charge of the large family. In 1799 she married Alexander Marcet, a Swiss physician who, at the time, held an appointment at Guy's Hospital. Her husband was an amateur chemist and a prominent member of London's scientific society (Dreifuss and Sigrist, 2012). Among their circle of friends were the chemist, J.J. Berzelius, and the feminist writer, Harriet Martineau.

This was the period of the great British chemists such as John Dalton, Joseph Priestley, and Humphry Davy. Davy gave a series of public lectures in London that were attended by members of the upper-class society. He was delighted by the large proportion of women in the audience, which included Jane Marcet (Golinski, 1999: 24). The book *Conversations on Chemistry* was written from first-hand experience. Having had difficulty understanding some of the lectures, Jane Marcet performed her own experiments. Finding that such a hands-on approach gave her a deeper understanding of chemistry, she decided to write an introductory textbook accompanied by experimental work so that others could comprehend the subject.

One of the great attractions of *Conversations on Chemistry* was the inclusion in subsequent editions of the latest discoveries that she learned from the famous British scientists, requiring the book to be expanded to two volumes. She was particularly keen to include the current research of Sir Humphry Davy, such as the isolation of the alkali metals, potassium and sodium. In the 1817 edition, the chapter 'Conversations on Metals' begins with this exchange:

> *Caroline*: I doubt, however, whether the metals will appear to us so interesting, ... they are bodies with which we are already so well acquainted.
>
> *Mrs. B.*: You are not aware, my dear, of the interesting discoveries which were a few years ago made by Sir H. Davy respecting this class of bodies. By the aid of the Voltaic battery, he has obtained from a variety of substances, metals before unknown, the properties of which are equally new and curious.
>
> (Marcet, 1817: 314–15)

The chemistry collection of the Edgeworths must have also contained Marcet's book. Reading *Conversations on Chemistry* had proved of particular importance to Edgeworth. Her chemical knowledge, acquired by reading Marcet's book, possibly saved the life of Edgeworth's younger sister: the sister had swallowed acid and Maria recalled from the text that milk of magnesia was an effective antidote (Armstrong, 1938: 55–6).

There were 15 British editions of Marcet's *Conversations on Chemistry* (the edition numbers went to the sixteenth, but the seventh was never published) totalling an amazing 20,000 copies in sales (Rossotti, 2006: v). *Conversations on Chemistry* was very popular in the United States, too, resulting in several plagiarized versions (Lindee, 1991: 16). In her account of the early teaching of chemistry in American girls' schools, Kim Tolley reported that Marcet's *Conversations on Chemistry* became widely used in the pioneering girls' establishments there (Tolley, 2003: 57–8). Unfortunately we have been unable to find any specific reference to Marcet's book in any of the British girls' schools we visited. Marcet died in 1858.

Chemistry at Quaker girls' schools

In the 1800s some of the Quaker schools were leaders in science teaching, and particularly in chemistry. The Act of Uniformity of 1662 had denied the Religious Society of Friends (Quakers) and other Dissenters the ability to found their own schools. However, even by 1691 at least 15 Quaker schools existed, and there were many more by the 1800s. As Geoffrey Cantor pointed out:

> However, if Dissenters could control the education of their young, they would be able to mould the attitudes and skills of ensuing generations and thus help to secure their denomination's future. Moreover, a high standard of education, especially in the sciences, was provided by many Dissenting academies.
>
> (Cantor, 2004: 147)

Many of the Quaker schools were designed for the children of Friends who were 'not in affluence' (Allen and MacKinnon, 1998: 398) and these schools emphasized a practical education useful for a future working-class life. A few Quaker schools, however, provided a more academic curriculum with higher aspirations of their graduates.

A fundamental belief of the Quakers was a spiritual equality of the sexes, so it followed that education for girls was as important as that for boys. Thus, at least in theory, girls' education should be equal to that of boys' (Stewart, 1953: 31, 66, 78). There were two specific reasons why Quaker women favoured an education for their daughters. First, the Quaker author Priscilla Wakefield, in her book *Reflections on the Present Condition of the Female Sex: With suggestions for its improvement* (Wakefield, 1798) wrote on the need for women to be educated in the context of marriage: '... an increase of real knowledge will conduce to give them [women] what they owe to themselves ... it will not teach them a servile, unqualified obedience, such as can only be observed by slaves ...' (ibid.: 86). In fact Wakefield and her friend, Maria Hack, specifically popularized science and promoted scientific literacy, especially for girls, in the early nineteenth century (Leach, 2006: 69–70).

Second, there was a dearth of Quaker men to marry, for example, at the York Monthly Meeting of Friends in 1840, 32 per cent of attendees were single women (Wright, 2011: 88). Being single was looked upon positively in the Quaker community, but the expectation was that an unmarried Quaker woman would contribute in some way to society, such as through teaching. Therefore, she herself needed to be educated if this was to be her lot in life.

In a history of The Mount School, York, Winifred Sturge and Theodora Clark commented in general on the reason why Quaker girls' schools benefited from the austere aspects of the Quaker movement:

> Looking back to the [eighteen] sixties, we may feel satisfied that among the girls' schools of the country, Quaker schools were indubitably in the front rank. The very narrowness of the curriculum – no music, dancing, or singing, no fine needlework – left space and time available for better grounding in history, arithmetic, and geography, and for relatively wider reading in English literature.
>
> (Sturge and Clark, 1931: 109–10)

But science was not just a time-filler. To Quakers there was a more fundamental link between their faith and science, in that science revealed the wonders and beauty of the universe and therefore gave them a deeper insight into God's works. At the Natural History sessions of the 1838 Newcastle meeting of the British Association for the Advancement of Science, it was reported that 'lots of ladies, mostly Quakeresses' (Owen, 1894: 126) were present. In fact, several influential Quaker women, such as Caroline Fox, had a fascination for chemistry (Rayner-Canham and Rayner-Canham, 2009: 120).

Although there were both co-educational and single-sex Quaker schools, we will focus on the girls-only schools. During the nineteenth century, there were numerous small Quaker girls' schools, many of which had short lifespans. As an example, the 1879 directory *Schools for Girls and Colleges for Women: A handbook of female education chiefly designed for the use of persons of the upper middle class* (Pascoe, 1879) carried an advertisement for a School for the Daughters of Friends and Others in Southport, run by Mr and Miss Wallis. The curriculum claimed to prepare pupils for the university examinations, including classes in science (ibid.: 62).

Chemistry was specifically part of the curriculum at Sarah and Harriet Hoare's Quaker school in Frenchay, near Bristol. Jane Heath, a student at the school, wrote in 1820 to her mother:

> We rise a little before seven and study Geography till eight with dissected maps. After breakfast we make our beds and into school again by nine ... I have then to write a page of English and Natural History, lectures on Chemistry, Botany, etc., besides parsing and a slateful of exercises so that I cannot always finish before dinner.
>
> (Heath, 1820)

Newington Academy for Girls

The Newington Green area, now a suburb of London, had always been a haven for religious dissenters, particularly Quakers. Thus it is not surprising that several girls' schools were established here (Baggs *et al.*, 1985). From 1668, there was a Quaker girls' school in nearby Shacklewell, run first by Mary Stott and then Jane Bullock, 'to Instruct younge lasses & maydens in whatsoever thinges was civill & useful in ye creation' (Penney, 1911: 119). And it was in Newington Green in 1785 that Mary Wollstonecraft, the pioneering feminist writer and philosopher (Todd, 2000) opened her boarding school for girls, following which she published *Thoughts on the Education of Daughters* (Wollstonecraft, 1787).

The first girls' school to note the teaching of chemistry was the (Quaker) Newington Academy for Girls, occupying half of Fleetwood House, Church Street, Newington Green. Among the founders and benefactors were the Quaker scientist William Allen and his third wife, Grizell Hoare. Allen's main occupation was as a pharmacist and manufacturing chemist but from 1802 he assisted William Babington with his chemistry lectures at Guy's Hospital. Allen also gave science lectures at the Royal Institution at Humphry Davy's request (Sherman, 1851: 74–5).

Opened in 1824, the Newington Academy for Girls started with 12 pupils, but this figure more than doubled in three years. The original prospectus stated:

> The course of Instruction shall comprehend a Grammatical knowledge of the English language, Writing, Arithmetic, Geography, Astronomy and the Use of Globes, Ancient and Modern History, Elements of Mathematics, of Physics or Experimental Philosophy, Chemistry, Natural History, The French Language and Needlework.
>
> (Shirren, 1951: 160)

It was Allen who taught chemistry, as his biographer noted: 'He [Allen] went through a course of lectures, which he annually repeated, on mechanics, chemistry, and natural and experimental philosophy, ... by familiar explanations, and by a variety of experiments with his extensive and valuable apparatus' (Sherman, 1851: 346). One of the students, Louisa Stewart, later recalled: 'The school was a pioneer of true high-class education ... William Allen gave the girls lessons in his own house in chemistry ...' (Roberts, 1970: 31–2).

In her memoirs, *Threescore Years and Ten,* Sophia Elizabeth de Morgan (née Frend) described how she had visited some of Allen's classes:

> ... I made acquaintance with William Allen, who kindly allowed me to attend the lectures on chemistry which he gave, with experiments, to a class of young girls. From him I learned the meaning and importance of Dalton's discovery. The atomic theory, then beginning to be understood, was the first step in the raising of chemistry to the rank of a science. Mr Allen's quick perception of facts was greater than his power of following out extensive inferences. He was a good observer and classifier, but stopped at facts and phenomena.
>
> (De Morgan, 1895, 106–7)

De Morgan, raised a Unitarian, had received no formal education. However, in her teenage years, she would sit in the branches of an old oak tree and self-educate. She commented: 'I used to climb this on fine mornings, and had read a very miscellaneous collection of books in the oak: ... books on natural philosophy and chemistry ...' (ibid.: 108). Though there is no evidence as to what she read, the period of the 1810s and 1820s would make it feasible for Marcet's tome to be one of her chemistry books.

Joseph Peace (who later became the first Quaker member of parliament) was a visitor to the Newington Academy for Girls in 1827. He wrote a poetic account, *A Yearly Meeting Epistle from Friend Joseph in London to his Cousin Anne in the Country,* of his astonishment at the wide curriculum, of which the first three verses are shown below:

> Dear Coz in my last (which I sent by Friend P ...)
> I shewed the advantage as well as renown
> That our body of Friends cannot fail to acquire
> By the Female Establishment 2 miles from Town
> Where the pupils imbibe such astounding variety
> Of stores intellectual – I solemnly vow
> Since the earliest days of the Quaker Society,
> Such achievements by girls were ne'er heard of till now.
> No science, no art, in their tribe is a mystery
> The path of the earth and the tides of the sea,
> Cosmography, Algebra, Chemistry, History,
> To those juvenile Blues are a mere A.B.C. ...
>
> (Shirren, 1951: 163)

From 1824 to 1838 the headmistress of the school was Susannah Corder, but the school then seems to have closed, to reopen briefly in 'The Laurels' on the other side of Church Street in 1841 under Sarah Sweetapple, and closing finally in 1848 (ibid.: 166).

The Mount School, York

It was in 1784 that a meeting of influential Friends was convened in York to consider establishment of a Quaker girls' school in Yorkshire. As a result of the support for the venture, a Quaker girls' boarding school was opened in York in 1785 under the supervision of Esther Tuke. However, the venture was not a success and the school closed in 1814 (Sheils, 2007: 19–27).

When a Quaker girls' school (later called The Mount School) was re-established in 1831, the senior mistress was Hannah Brady (née Smith), aged 28. Even before she was 21 Brady had been a teacher on the girls' side at the co-educational Quaker school at Ackworth, Yorkshire. Her husband, Henry Brady, a master at Ackworth, had died within a year of their marriage.

Chemistry at The Mount School

It was Hannah Brady who first encouraged an interest in chemistry among The Mount girls. The evidence comes from the letters written in 1835 by the schoolgirl Anne White which were reprinted in Sturge and Clark's school history, *The Mount School, York*:

> Well we went yesterday evening to the boys' school [Bootham Quaker boys' school]. H[annah] Brady came with us and we took tea with the brothers and cousins of the girls … After tea we looked at pictures until John Ford was ready to show us some chemical experiments. We then went to the dining-room where we found 10 or 12 more boys sitting and were shown a great many more experiments, most of which I saw before, but I was very glad to see them again. John Ford showed us phosphorus and sulphur burning in oxygen, potassium jumping about in water, phosphorus dissolved in water and boiled in a flask, several striking affinities, particularly making plaster of Paris, which I do not think I ever saw before … I spent the pleasantest evening I have had since I came.
>
> (White, 1931: 43–4)

Anne was obviously not the only girl enthralled, as she added:

> H[annah] Brady and the older girls are very much interested about Chemistry and Natural Philosophy. I wish Dada [her

father, James White, a Quaker schoolmaster of Ballitore, Ireland]
could tell some simple experiments we could do without much
trouble. I find the little knowledge I have of it of great use to me.
I believe it was for the experiments we were invited to the boys'
school. Whatever it was for, all the girls agree that they spent a
delightful evening.

(ibid.: 47)

According to the Mount School history by Mary Smith and Elizabeth Waller:
'... mathematics and science [was first taught at The Mount School] by an
indolent Irishman, and later, in striking contrast, by Edward Grubb ... His
lessons and also his contributions in Sunday morning Meeting were thrilling
and adventurous, opening new horizons to many' (Smith and Waller, 1957:
10). Sturge and Clark also commented:

To his lectures [Edward Grubb's] on chemistry his audience came
in a mood of prophetic sympathy, awaiting the experiment: "Will
it? Won't it?" It generally wouldn't! Why should it? For before
the laboratory was built in 1884 there was no science equipment
worth the name.

(Sturge and Clark, 1931: 131)

Grubb taught chemistry at The Mount from about 1882 until 1892 (Dudley,
1946: 46–7). Then from 1892 until 1895 he taught chemistry at Polam Hall
(see below) (ibid.: 59). In 1895 Grubb received an offer of a position at
Brighthelmston School, Southport, a girls' boarding school, and he and his
family settled there (ibid.: 60).

From 'The Mount Magazine'
As we discuss in more detail in subsequent chapters, in the late nineteenth
and early twentieth centuries, the school magazines provided a valuable
insight into the girls' attitude to chemistry and their knowledge-base, and
The Mount Magazine was no exception. Students at many schools in the
time frame of our study wrote alphabet songs, sometimes with reference to
chemistry in one or more lines. This was true at the Mount School, where
there were two such references:

[Alphabet]
L is for "Lab" – meant for people with brains; where are crystals
& acids, explosions & stains.

(Anon., 1913: 95)

Mount Alphabet

L's for the **Lab** which is full of queer smells.

(Anon., 1920a: 113)

Another favoured format for simple rhymes at girls' schools was the song 'Ten Little Indians'. At The Mount School, one of the verses had a chemistry emphasis:

Five Matriculation girls behind the Lab door
One sniffed the Bromine jar & then there were four.

(Anon., 1898a: 43)

'O Tannenbaum' was the basis of the following rhyme in 1893. The emphasis on the properties of gases suggests that qualitative analysis, involving the production and identification of gases, was a significant part of the laboratory work. In those days – and even in more recent times – it was regarded as quite natural to make and drink afternoon tea in the lab:

Oh charming lab oh pleasant lab
When I enter through thy door
I see thee smeared with many a dab
Of H_2SO_4
Solutions, solids, crystallines
Upon thy shelves do stand
With acids, bases, alkalis
Forming a glorious band
Oh many are thy odours
And pleasant ones but few
But worse than all the others
Is that of SO_2
Often upon thy bench is seen
A kettle, cup, or spoon
And many a brew of tea I ween
Is made each afternoon
And in thy corner cupboard
Are things we can but guess
They are not so apparent
As fumes of H_2S
...

31

Oh charming lab! Oh pleasant lab!
I oft shall think of thee
In the dim and distant future
And the days that are to be.

('Vera', 1893: n.pag.)

A unique artistic endeavour at The Mount School was to write a chemical analysis table in which the ion being tested is replaced by an academic subject and a reference to the textbook for that subject:

Add Industry; ppt. indicates	Digest ppt. time	Confirmatory Test
LITERATURE	Insoluble	(i) Add dilute Masson; gas evolved which turns exam: paper black. (ii) Add Prose periods in excess; thick ppt. of pamphlets.
FRENCH	Partially Soluble	(i) Heat with "Brachet". & evaporate, fragments of rules left on sides of test-tube. (ii) Add "Dictée" much ppt. of mistake.
ENGLISH LANGUAGE	Soluble	Add dry Grimm & Verner & heat with questions; **A** appears (this result seldom obtained)
MATHS	Easily soluble	Add brains if obtainable; otherwise notes. Answers given off. (N.B. this test needs care in working)
LATIN	Partially soluble	Add unseen; & cook; gas of Sense <u>sometimes</u> evolved.

(Anon., 1895: n.pag.)

And, equally original, a page in *The Mount Magazine* (Figure 2.1), showed pen-and-ink sketches of animated chemistry laboratory equipment with fanciful names (Anon., 1898b: 59).

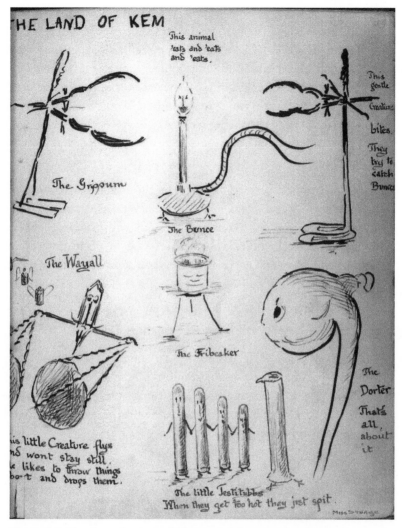

Figure 2.1: Sketches of animated chemistry laboratory equipment
Source: *The Mount Magazine*, 1898b: 59

In addition to classes and laboratory work on chemistry, the girls were taken to lectures in York relating to chemistry. For example in 1897 in *The Mount Magazine* under 'School Events', we read that:

> <u>Thursday, March 11th</u>. First and Second Classes went to hear an Albert Library Lecture by Theodore Field on John Dalton, the Quaker Chemist. The balls by which Dalton illustrated the Atomic Theory & other relics of interest were shown.
>
> (Anon., 1897: 2)

Then it was reported in *The Mount Magazine* that on 8 February 1915 the girls had attended the first chemistry lecture at the Tempest Anderson Hall (Anon., 1915: 10). There had been a regular lecture series on science organized by the Yorkshire Philosophical (what we would now call Scientific) Society since 1822. In 1912 the hall was opened as the new venue of the lectures, and this event attended by The Mount students was the first on the subject of chemistry.

The formation of a combined Scientific Society at the school was announced in 1927, but membership was a privilege:

> … Probationers become members on the attainment of a certain number of marks given, at the discretion of the committee, for work done. This work must include some satisfactory records of original scientific observations. There will be a Scientific ribbon for members who have reached a certain standard in their work.
>
> (Booth, 1927–8: 100)

The following year, the Scientific Society was active demonstrating experiments in the laboratory during Exhibition Day:

> … Soap was being made by A. Linday, who had been experimenting on this subject most of the term. The various stages in the manufacture of beet sugar were shown, also the process of making bath salts, the distillation of tangerine oil, and various other experiments …
>
> (Anon., 1928–9: 94)

Polam Hall, Darlington

Founded in Darlington in 1847, Polam Hall was a Quaker girls' boarding school (Davies, 1981). A very early prospectus exists for *Jane, Barbara and Elizabeth Procter's Boarding School for the Daughters of Friends* (Figure 2.2) which states that: 'Lectures on Natural History, Natural Philosophy, Chemistry, &c., are regularly delivered to the Pupils' (Archives, Polam Hall). Unfortunately, the document is undated, but Barbara Procter died in 1851, which would suggest the prospectus was issued prior to that date. There are no details about the teaching of chemistry except that Samuel Hare and Edmund Wheeler are listed as teachers of scientific subjects during the 1880s when the Procter sisters still ran the school (Archives, Polam Hall).

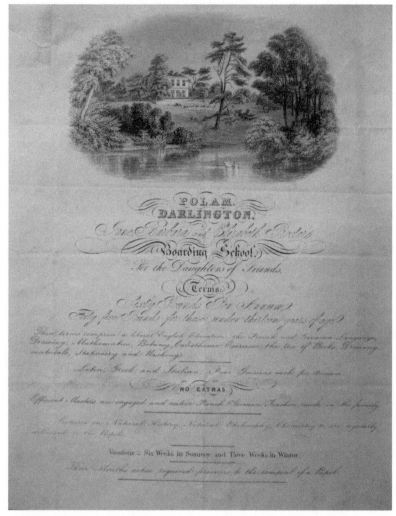

Figure 2.2: The early prospectus for Polam Hall, stating that chemistry was among the subjects taught

Source: Archives, Polam Hall

Also in the Polam Hall archives and undated is an aged handwritten timetable of a student. Chemistry is shown for Thursday, 10:50 a.m. to 11:30 a.m. and Friday from 5:30 p.m. to 6:50 p.m. The timetable was most likely from the 1892 to 1895 period when Edward Grubb taught chemistry at Polam Hall.

During the early 1900s classes of senior girls attended the Darlington Technical College for Science for classes with A.F. Hogg. In 1907 a science club was founded and it was reported that it was started:

> ... to encourage girls to work independently at some branch of science which interested them. Most excellent exhibitions of individual collections took place every year, and contributed greatly to the very high standard which the Science work of the school reached ...
>
> (Archives, Polam Hall)

The first individual identified as a qualified chemistry teacher at the school was Robert H. Sargent. From 1911 until 1934 Sargent, a science lecturer at the Darlington Technical College for Science, visited Polam Hall to teach chemistry. In an obituary for Sargent by Helen Baynes in the school magazine, *By Kent and Skerne,* she mentions that Sargent:

> ... helped us make a [chemistry] laboratory out of a basement room, No. 29 at Polam House, with basins, gas, office stools and tables, and much good work was done there. It is an example of what we all know that, in teaching, the essential thing is not the equipment (valuable as that is), but the personality and skill of the teacher.
>
> (Baynes, 1940: 6)

A proper chemistry laboratory was finally constructed in 1920 (Figure 2.3), as an enthralled student wrote in the school magazine:

> We were very excited when we came back after 'Xmas to find Olympus [presumably the students' name for the previous use as an attic dormitory] transformed into a wonderful new laboratory. The new "lab" had been spoken about for so long that we had almost given up hope of ever having it, but here it was at last – the old room scarcely recognizable with all its new fittings ... We are all very proud of it and were very anxious to show it off at Old Scholars' ...
>
> (Anon., 1920b: 4)

Figure 2.3: The 1920s chemistry laboratory at Polam Hall. In the right foreground can be seen the apparatus for collecting gases over water
Source: Archives, Polam Hall

Finally in 1931 with the arrival of Miss Armitage, Polam Hall had its own full-time chemistry teacher (Archives, Polam Hall).

Although the Quaker girls' schools were the earliest to offer chemistry, they were outside the mainstream of education. The few Quaker girls' schools were both geographically and culturally isolated from the rapid growth of science-active academic girls' schools during the 1880s. For the commencement of this narrative, we need to focus on two role-model schools: North London Collegiate School and Cheltenham Ladies' College and their charismatic heads, Frances Buss and Dorothea Beale.

Chapter 3

Chemistry and the two role-model girls' schools

...

"The time is come," Minerva said,
  "For you, the school's élite,
To study Greek and Chemistry,
  Acoustics, Light, and Heat;
to talk of that mysterious sphere
  Where parallels all meet."

...

('Two Little Oysters', 1879: 12)

By 1879 chemistry was well established at North London Collegiate School for Girls (NLCS) as can be seen from the above extract of the girls' parody of the Walrus and the Carpenter's rhyme in *Alice in Wonderland*. In the Taunton Report of 1868 (see chapter 1), two independent girls' schools were singled out as exemplifying the best in academic programmes for girls. One of these was the day school NLCS under Frances Buss and the other was the boarding school Cheltenham Ladies' College (CLC) under Dorothea Beale. In fact these two principals were so highly regarded that they were requested to give testimony at the Taunton Commission hearings (Kamm, 1958: 65–98). This chapter, then, is about chemistry at these two model schools. We must begin with a brief background on the two principals and how their educational views were moulded.

Frances Buss and Dorothea Beale

"Miss Buss and Miss Beale
Cupid's darts do not feel:
How different from us,
Are Miss Beale and Miss Buss."

('Paedagogus Emeritus', 1907: 75)

This rhyme (the original version being given here), composed over 100 years ago, remains in circulation in academic papers (Davison, 2009) and even as quiz clues (Anon., 2014). However, few may know the identity or work of Frances Buss and Dorothea Beale. Yet it is fitting that their names are still recalled – and together – as exemplified by the title of Josephine Kamm's book *How Different from Us: A biography of Miss Buss and Miss Beale*, for the two of them fundamentally changed the nature of educational opportunities for middle- and upper-class girls (Kamm, 1958). Their goals of girls' academic education were similar, though the schools that they developed were significantly different. One striking commonality in their backgrounds was their own educational experiences at Queen's College, Harley Street.

Queen's College, Harley Street

In the early history of the academic education of women, Queen's College in Harley Street, London, played a crucial role. It was in 1847 that a group of professors of King's College, London, started a series of 'Lectures to Ladies'. The intended audience for these lectures were governesses. During the middle of the nineteenth century, the position of governess was the only employment option for unmarried middle-class women and, at the time, few of them had any formal education. These classes were so successful that they provided the springboard for the establishment of Queen's College in 1848, the first institution for the comprehensive academic education of women (Grylls, 1948; Kaye, 1972).

Queen's College was founded by Frederick Denison Maurice, a Christian Socialist. The Christian Socialists believed that religion and social progress went hand-in-hand, and that, in particular, women had rights to education. The college was run by men, thus it was deemed essential that an adult woman be present in each classroom to chaperone the proceedings (Spencer, 2006). The title 'Lady Visitors' was used for these individuals, one of whom was Jane Marcet (see chapter 2). Maurice obviously thought very highly of Marcet as in his discussion of Lady Visitors for Queen's College, he footnoted: 'Among these ladies we have the high honour and privilege of reckoning one whose life has been devoted to earnest and successful efforts for the instruction of both sexes and all ages – Mrs Marcet' (Maurice, 1898: 17).

In her analysis of the Queen's College curricula from 1848 to 1868, Shirley Gordon noted:

> The scientific studies at Queen's College were, for the period,
> amongst the most remarkable in the curriculum. At a time
> when it was very much an open question whether boys should

study science, the committee at Queen's College included Natural Philosophy in the survey of academic knowledge which represented their timetable ... The introductory syllabus in Natural Philosophy, outlined in 1854, is interesting ... Thirdly, the class was to hear of the leading facts and classifications of chemistry and to see experiments to demonstrate the points.

(Gordon, 1955: 150–1)

One of the students of that time, Maud Beerbohm Tree, recalled: 'At 12:30 I think we descend into cavernous depths of the earth to that quaint home of Natural Philosophy, Chemistry, and Mathematics known as Mr Cock's room' (Tree, 1898: 44). Later in the nineteenth century, chemistry was taught as a separate subject, first by William A. Miller, next by J. Millar Thomson, and then in 1897 by H. Forster Morley, all chemistry professors of King's College, London. The Chemistry Examiner for 1897 was listed as Richard T. Plimpton, a chemistry professor of University College, London (Croudace, 1898: 30–2).

Frances Buss

It was arguably Frances Buss's experience at Queen's College that led her to include a very strong science programme when she subsequently founded the NLCS. Buss was born in 1827, daughter of Robert William Buss, painter and etcher, and Frances Fleetwood. Initially, she was sent to a small private school in Aldersgate, then to one in Kentish Town, and finally, at age ten to Mrs Wyand's School in Mornington Place. By the age of 14 Buss was teaching at Mrs Wyand's School and by 16 she was sometimes left in sole charge of the school for a week at a time (Ridley, 2012).

To help with the family finances, in 1845 Buss's mother set up a private school for young children in Clarence Road, Kentish Town. The methodology of the Swiss educationalist Johann Heinrich Pestalozzi was used at the school, that is, teaching was child-centred and based on individual differences, sense perception, and the student's self-activity. Buss and the other family members assisted teaching at the school, which they subsequently moved to a larger house on Camden Street.

Buss was painfully aware of her own limited education. Thus during the 1848/9 academic year she attended evening lectures at Queen's College, Harley Street. Many years later in a letter to Dorothea Beale, Buss described the education she had gained there as opening: 'a new life to me, I mean intellectually' (Buss, 1889). Attending Queen's College was a considerable challenge for there was no public transportation and Buss had to walk from

Camden Town to Harley Street in time for the evening classes, then back again late at night, and study into the early hours of the morning.

In 1850 Buss took over her mother's School as headmistress, renaming it the North London Collegiate School for Ladies. She was only 23 years of age.

Dorothea Beale

Dorothea Beale, born in 1831, was the daughter of Miles Beale, surgeon, and Dorothea Complin (Shillito, 2010). She was initially educated by a governess but then was sent away to school – an experience that convinced her of the need for a radical overhaul of girls' education. Beale's autobiographical notes were compiled by Elizabeth Raikes, who commented:

> 'It was a school,' again to quote Miss Beale's own account of her education, 'considered much above average for sound instruction; our mistresses had taken pains to arrange various schemes of knowledge: yet what miserable teaching we had in so many subjects; history was learned by committing to memory little manuals; rules of arithmetic were taught, but the principles were never explained.'
>
> (Raikes, 1908: 9)

Beale had to withdraw from formal schooling at the age of 13 due to ill health. This was a 'blessing in disguise' for through her avid reading at two large London libraries, and with her father a member of the Medical Book Club, she learned far more than at the poor-quality girls' schools of the time.

Entering Queen's College, Harley Street, when it opened in 1848, Beale so impressed the staff that after completing her studies, she was appointed mathematics tutor. However, with the arrival of a new principal, Dr Plumptre, Beale became increasingly disillusioned. In part, it was because she believed standards were being lowered and students were being admitted without sufficient ability. The other major factor was that women tutors had less and less authority under Plumptre. In 1856, when women tutors became even more marginalized, Beale decided the time had come to resign.

Beale then accepted a position as headteacher at the Clergy Daughters School at Casterton, Westmorland, a school for the daughters of impoverished clergymen. Her attempts to bring that school up to her vision of an academic institution for girls failed and she left within a year. During the next year, Beale wrote *The Student's Textbook of English and General History* (Beale, 1858) and the success of this book contributed to her appointment as second principal of CLC in 1858. Here, at last, she was able to put her strong principles into action.

Originally named the Young Ladies' College, Cheltenham, the CLC had opened in 1854 (Clarke, 1953). Over the first four years, the enrolment declined. When the Governors appointed Dorothea Beale as the second principal in 1858, everything changed. Academic education became the focus, using the boys' Cheltenham College as a model. The construction of science facilities was one of Beale's priorities. However, she had to innovate slowly in such a conservative society as that of Cheltenham. She recalled that in the early years, a father had withdrawn his daughters from the school as he objected to them learning arithmetic (Beale, 1880: 58). To avoid any additional friction with the parents, chemistry was initially introduced under the guise of 'physical geography and science'. Beale remained as Principal of CLC until her death in 1906 – a 'reign' of 48 years.

A difference between the NLCS and CLC models

Though Buss and Beale shared an enthusiasm for girls' academic education, there were several ways in which their views differed significantly. One of these was in the social strata of the girls they planned to attract. Beale's view was expounded at the Taunton Commission hearings. She was asked whether her pupils were of the upper portion of the middle class. Her response was that none were admitted other than 'daughters of independent gentlemen or professional men' (Kamm, 1958: 82).

By contrast, in the early years, NLCS was viewed as a day school for the education of the daughters of the local middle-class community, often of limited means, such as the daughters of clerks or tradesmen (Steinbach, 2004: 174–5). One of the students, Molly Hughes, had been delighted to move from a fashionable 'snobby' school to NLCS: 'Now at the North London I sensed at once a different atmosphere. No one asked where you lived, how much pocket-money you had, or what your father was – he might be a bishop or a rat-catcher' (Hughes, 1936: 50).

Many of the other British girls' boarding schools were modelled upon the CLC. With their high fees and somewhat exclusive clientele, any contact with slightly inferior ranks of society could be avoided. One example was Downe House. A former student was quoted in the school's history as recalling:

> 'My own mother,' writes one old Downe House girl who went to the school in 1914, 'was not a snobby person but she would not send any of us four girls to the excellent education of the Oxford High because she was afraid we would "pick up accents",

and it would be awkward if we became friends with butchers' daughters.'

(Ridler, 1967: 77)

On the other hand, Oxford High School, mentioned in the quotation above, as a school of the Girls' Public Day School Company (GPDSC), was modelled upon the NLCS. A student who had attended Notting Hill High School, another GPDSC school, recalled:

> The girls were of all social conditions in my time. I sat between the daughter of a publican and the daughter of a laundress, and I never succeeded in beating the former. The daughter of a viscount was at the bottom of the class.

(Swanwick, 1935: 75)

Chemistry at North London Collegiate School

Buss wanted her school to give girls an academic education: she wanted her pupils to have equal opportunities in common with boys and that included real science. Her father, Robert Buss, had educated himself so that, initially, he could teach the science classes. In a history of the school, Nigel Watson remarked:

> It [NLCS] also set out to offer their daughters an education quite different from anything available anywhere else. There was, for example, the inclusion of science within the curriculum. Science was not really taken seriously by most girls' schools until well into the twentieth century. Robert Buss [Frances Buss's father] made a memorable science teacher as Annie Martinelli, an early pupil later remembered: 'His talents were simply wonderful ... His Chemistry series was marvellous, especially for smells and explosions.'

(Watson, 2000: 16)

In 1875, with the construction of a new school building on Sandall Road under way, Buss lobbied strongly for a proper chemistry laboratory to be added. The finance and construction of this addition was approved by the Charity Commissioners (Charity Commission, 1875). The laboratory (Figure 3.1), one of the most advanced of its time, was designed by the renowned architect of technical and college buildings, Edward Cookworthy Robins. The original plan for the NLCS laboratory was included in his monograph *Technical School and College Building* (Robins, 1887: plate 52).

This outstanding facility enabled the students to do hands-on experiments themselves. An article on NLCS by Mrs Roscoe Mullins that appeared in an 1893 issue of the girls' magazine, *Sylvia's Journal*, showed the students at work in the laboratory (Mullins, 1893).

Figure 3.1: Chemistry laboratory at the North London Collegiate School, Sandall Road, built in 1879

Source: Mullins, 1893: 503

During her time at NLCS in the early 1880s, Molly Hughes recalled a laboratory experience with her friend, Mary Wood:

> At school we had both indeed dabbled in a little chemistry in the lab., with bunsen burners and sinks and test tubes, but our enthusiasm cooled after Bessie Davis let loose the H_2S on us one day, and little remained to us but a few choice words, which we employed, like Humpty Dumpty, to mean just what we wanted them to.
>
> (Hughes, 1936: 177)

Headmistress until the end, Buss died in 1894. The strong science programme at NLCS was continued by the second headmistress, Dr Sophia (Sophie) Bryant (née Willock). Born in 1850, Willock had been home-schooled by her father until she entered Bedford College (then more of a girls' high school) in 1866. She married Dr W. Hicks Bryant in 1869, and upon his death in 1870, obtained a teaching position at a school for ladies in Highgate. In 1875, she joined the staff at NLCS to teach mathematics. Prior to her death,

Buss had recommended Bryant as her successor, a position to which Bryant was appointed in 1895.

Under Bryant, a large extension of the school was constructed in 1909, including a new chemistry laboratory (Figure 3.2). The laboratory, which would have been considered modern even in the late twentieth century, was described by Eleanor Hill in *Our Magazine: North London Collegiate School for Girls*:

> At one end is a blackboard which can be raised. Behind this is a large fume cupboard fitted with gas, water, and electric light, while on each side of the black-board is a small fume cupboard … Each bench has three sinks which can be used as pneumatic troughs, and there are three taps to a sink … There are balance benches on each side of the room, and another bench on the window side fitted with gas and water. There is also a still for preparing our own distilled water, and the demonstration table is fitted with two sinks, gas, water, and electric light … Just beyond this room is a store room for all Science apparatus.
>
> (Hill, 1909: 48)

Figure 3.2: The second chemistry laboratory at the NLCS

Source: Archives, NLCS

Chemistry teachers

An excellent laboratory facility was only part of the equation: having outstanding chemistry teachers was another necessity, and both Buss and

Bryant ensured this was the case. The first woman chemistry teacher was Dilys Davies (NLCS, 1884). Davies, born in 1857, had been a student at NLCS (Evans, 2014). After spending the 1877/8 year at Newnham College, Cambridge, she returned to NLCS to teach chemistry from 1879 until 1888. Following her marriage in 1889 and a move to Bangor, Davies (Mrs Dilys Glynne Jones) dedicated the rest of her life to the cause of girls' education in Wales (see chapter 9).

As the first qualified woman chemistry teacher, Buss hired Annie Grace Heath in 1888. Heath had been a chemistry student with Henry Armstrong (Eyre, 1958: 272) at the Central Technical College. Heath, a firm believer in Armstrong's heuristic method of teaching chemistry (see chapter 4), had a passion for the subject. One of her former students, Dora Bunting, wrote:

> ... Chemistry was the subject she devoted most of her time to. She was enthusiastic about it, and tried to make her pupils so, to develop their faculties of thinking and working out everything for themselves. This was especially so in the practical work in the laboratory. She tried to insist on each one doing the work for herself, and understanding each experiment and the reason for it, so that we should not be content with learning what the results should be.
>
> (Bunting, 1895: 29)

Sadly, in 1895 Heath contracted pulmonary tuberculosis. Her successor, Edith Aitken, visited Heath just before she died: 'She [Heath] sat there so weak yet still so bright, and talked of Dr Armstrong's chemistry class ...' (Aitken, 1895: 27).

Aitken, born in York in 1861 (Megson, 2006), was sent at the age of 14 as a boarder to NLCS where her elder sister was already teaching classics. Attending Girton College, Cambridge, from 1879 to 1892, Aitken achieved first class honours in natural science in part one of the Tripos. As one of the 'steamboat ladies' (see chapter 5), she received an MA from Trinity College, Dublin in 1902. Before being hired at the NLCS, Aitken held teaching positions at Manchester High School for Girls (1882–3); Nottingham Girls' High School (1884–7); and Notting Hill High School (1887–92) – the latter two were GPDSC schools. Aitken, like Heath, proselytized for the heuristic method of teaching chemistry. Aitken left in 1902 to take up a position as head of the High School for Girls, Pretoria, South Africa. She died in 1941 ('N.F.W.', 1941: 725–6).

It was Buss's next recruit, Rose Stern, who was to give the teaching of chemistry at NLCS a high visibility for the next 28 years. Born in 1869 in Birmingham, Stern was the daughter of Moritz Stern and Fanny Schwartz. Graduating in 1889 from King Edward VI High School for Girls, Birmingham, Stern completed a BSc (London) from Mason College (later the University of Birmingham) in 1894. While at Mason College she had been the first woman Student Member of the Institute of Chemistry (Anon., 1893). Stern studied towards a teaching diploma at Cheltenham Ladies' College from 1895–6 (Staff Records, NLCS). From 1897 until moving to NLCS in 1902 she was science mistress at the Bangor County School for Girls (Anon., 1902: 46). Dilys Davies was on the governing body of the school in that period and she may have played a role in Stern's subsequent appointment at NLCS.

Figure 3.3: Miss Stern

Source: *The Searchlight*, Archives, NLCS

Stern, too, was a fervent believer in the heuristic method. But she put her own imprint on the chemistry teaching. As one student later recalled:

> "Stern by name and stern by nature" were the words in which she introduced herself to me. Luckily Miss Stern also had a sense of humour and a twinkle in her eye for many sinners. Although she was a merciless judge of slackness and inefficiency, she had a

very real patience with the hard working but not so bright pupil. She taught her classes to be independent and nothing pleased her more than to find that they were really experimenting and carrying on without too much help.

('K.N.H.H.', 1954: 44)

After retirement, Stern remained active in science teaching. For example, in 1932 she visited St Martin-in-the-Fields High School for Girls to give a presentation 'Faraday and his Work' (Dubois, 1932: 43). As a result of breaking a leg during the Second World War, she became less and less mobile, and died in October 1953 (Anon., 1954: 43).

The science club

The NLCS science club was founded in 1890 (Heath, 1890: 42–3). The first student presentation to the group was on the topic of Asbestos: '… the reader illustrating her remarks by numerous specimens, some of which she afterwards presented to the Club's Museum …' (Meakin, 1890: 92).

An annual event of the NLCS science club was the laboratory tea. This first took place in 1894: 'Miss Aitken and the 1894 Science Class … gave a Laboratory tea, at which tea was drunk from beakers and cake was eaten on porcelain dishes' (Blyth, 1894: 146). The event was next mentioned in 1903 by Isabel Soar, where it was noted that former pupils from the sciences had been invited back for the event 'and enjoyed an hour of pleasant social intercourse, added to which was the charm of drinking tea in beakers, and performing the difficult feat of eating ices out of crystallizing dishes with modified glass rods for spoons' (Soar, 1903: 103).

The tea of 1912 was very thoroughly documented. The invitations were sent to the staff on circles of filter paper with invisible ink and the words 'please iron' in normal ink. For their reply, one teacher sent back an apparent blank piece of paper with the single word 'iodine'. The students heated some iodine crystals and held the paper in the vapour, resulting in the answer becoming visible ('Editor', 1912b: 56–7). The last report of a laboratory tea that we could find was noted by Dorothy Adams in 1920 (Adams, 1920: 40–1).

The highlights of the science club meetings were visits to industrial chemical works. The first such adventure was with Grace Heath to Daintree and Company's Dye Works at Southwark, as chronicled by Edith E. Humphrey (see online Appendix 3):

We saw a vat full of *green* indigo, but we were told that the cloth which was then in it would turn blue when brought under the action of light. We also saw a red dye made by boiling logwood, which was mixed with a soluble ferric salt to produce black. We had specimens of the different aniline dyes given to us, and they may now be seen in our Museum.

(Humphrey, 1892: 142)

What comes through strongly in the later 'expeditions' is the adventurousness – and actually fearlessness – of the girls as they clambered around very dangerous environments in their long skirts. For example, in 1908, Rose Stern took a dozen senior science students to Messrs Pearce and Sons' sulphuric acid works at South Bromley:

First of all we were shown the pyrites burners, the sulphur dioxide from the burning iron pyrites being conveyed away by flues to the "chambers." ... We next saw the nitre pots, for the production of nitric acid, which is used as the source of the oxides of nitrogen in the acid process.

It was a very dark day, and no lights were used, so that from this point onwards we had a most thrilling time. We climbed up some very steep, very narrow, and very slippery stairs, and found ourselves in a dark passage, filled with fumes and very hot ... The windows in this passage were very few and far between, and the continuous drip-drip of the liquids inside the [lead] chambers, together with the holes in the flooring, which acted as sky-lights, added greatly to the weirdness of the whole. Here we were instructed as to the various reactions taking place inside the chambers, the sulphur dioxide, oxygen, oxides of nitrogen, and steam, ultimately producing sulphuric acid ... We then left the lead chambers, and, with assistance in front and encouragement behind, crossed a very unstable gangway, suspended over the pyrites burners ... Our descent was down steps more steep, more narrow, and more slippery than those of our ascent.

(Hewer, 1908: 9–10)

Then in 1929 Stern took the fifth form chemistry students to the South Metropolitan Gas Works. Not only was this another 'thrilling' adventure, it is clear that, as with the visit to the sulphuric acid plant, the girls knew their principles of industrial chemistry:

We were first taken to see the coal being burnt in the furnaces. When the doors were opened great flames rushed out and the heat became almost unbearable. The coke was then shot out into water and the gas conducted upwards and then cooled in pipes. On emerging from the shed we stared aghast at each other's countenances, streaked with coal dust ... We all feel that this visit was very thrilling and instructive.

(Anon., 1929: 82)

The science club magazine

Unique among the science clubs at the independent girls' schools that we visited, the NLCS science club had its own magazine. The first mention of the magazine was in 1891 by Melicent Wilson: 'The first volume of *Our Science Club Magazine* was started on its rounds last Term, and we hope that the second volume will soon be ready' (Wilson, 1891: 42). In 1912 the magazine was renamed *The Searchlight: NLCS Student Magazine for Science,* and the issues of 1912, 1913, and 1914 have survived.

The first issue of the *Searchlight* contained the description of an experiment that some of the senior students performed, entitled 'Another Experiment':

Object.   To do something exciting.

Experiment.   We placed a considerable quantity of potassium chlorate & crushed sugar in a trough & poured onto it a little concentrated sulphuric acid.

Results.   1. The mass burst into flames.
2. Miss Stern was much alarmed & required a lengthy explanation, but even after she had received this, would only concede that the brilliant experiment <u>might</u> have been a success if we had used smaller quantities.

Conclusion.   Experiments which may be thought in any way exciting should <u>not</u> be performed when nervous people are about.

('Chief Experimenter', 1912: 26)

As we explore more in chapter 8, many of the chemistry students expressed themselves in verse. In the pages of the *Searchlight* were numerous attempts at rhyme. One such effort by Florence Silk was based on the religious poem that starts: 'I am a Little Soldier' (Lawrence, 1871: 58). Of the nine verses, we have included verses two and seven here:

I am a little molecule
I contain some SO_4
If you are not careful
I'll burn your pinafore   sulphuric acid

I am a little molecule
Nitrogen atoms two
If you do inhale me
You'll laugh like a cockatoo   nitrous oxide

(Silk, 1913: 6–7)

Another student, Doris Allen, bemoaned her hours of studying chemistry, adding coloured sketches of herself to match each verse (Figure 3.4). Here we have included the first two verses of seven:

For many a weary hour or twain,
I've wrestled with my chemistry,
The doughty Newth doth cause me pain –
E'en Shenstone I peruse in vain!
Appalling is my destiny

Figure 3.4: Illustration to accompany the first verse of the poem by Doris Allen

Source: *The Searchlight*, Archives, NLCS

I prate of formulae & know
The pleasant smell of H_2S
That phosphorus a light will show,
That CO fumes can lay one low,
That's all I must confess

(Allen, 1913: 31)

The mention of Newth refers to the textbook, *A Text-Book of Inorganic Chemistry* (Newth, 1894), a classic of the time; likewise, Shenstone refers to Shenstone's textbook, *The Elements of Inorganic Chemistry for Use in Schools and Colleges* (Shenstone, 1900).

One challenge faced by the students was a three-hour practical chemistry exam. The anticipated disasters were similarly put in verse, these two stanzas being typical:

When do the burettes always leak
The test-tubes smash, the glass rods squeak?
Though you feel so very meek?
In three hours Practical!

When does the acid always pour
Itself right down your pinafore
Of blue with rents & holes galore?
In three hours Practical!

('Sub-Editor', 1913: 17)

An outstanding watercolour by Muriel Sutton was included in the 1914 issue (Figure 3.5). This illustration (Sutton, 1914: frontispiece) depicted the test used to identify borates in an unknown sample. The boric acid is dissolved in methanol and then concentrated sulphuric acid is added to produce volatile trimethyl borate which can be ignited to give a green flame.

As in the opening to this chapter, rhymes were often parodies of established poems. The following poem is clearly based upon 'Hail to thee, blithe Spirit! / Bird thou never wert', the first two lines of *To a Skylark* by Percy Bysshe Shelley. Students of the period would have used *Poems That Every Child Should Know*, which contained this particular poem (Burt, 1904). In this version, for which we have included the first two verses of seven, the unsuccessful test for ozone was performed to investigate the published claim of the presence of ozone on the Central Line of the London Underground. This test involved exposure of a starch-iodide test strip to the atmosphere, where presence of ozone would have turned the paper blue:

THE BORACIC ACID FLAME

Figure 3.5: Watercolour art based upon the green flame from the boric acid test

Source: Sutton, 1914: frontispiece,
The Searchlight, Archives, NLCS

Ode to Ozone

Hail to thee pure ozone!
In the tube thou'rt not,
As we've with papers shown.
On them thou mads't no spot,
No spot of <u>palest</u> blue, not e'en one tiny blot

The deep blue we had hoped for
Never did appear!
We'll seek thee no more,

For thou art not there!
And, trusting the control, no trace of thee is here.

('Editor', 1912a: 29)

Chemistry at Cheltenham Ladies' College

The teaching of chemistry at CLC dates back to at least 1875. In that year, C.L. Bloxham, professor of chemistry at King's College, London, administered an examination in experimental science. Among the questions were:

> What chemical compound is that which contains 14 parts by weight of nitrogen, 48 parts of oxygen, and 1 part of hydrogen?

> A lighted taper is introduced into jars containing different gases; in A it is simply extinguished; in B it burns very brightly; in C it is extinguished, but kindles the gas, which burns with a distinct blue flame; in D the same effects are observed, but the flame is scarcely visible in broad day. What may the gases be?

> (Anon., 1875: 56)

In the context of science, the first of Beale's changes was the conversion in 1876 of the calisthenics room to a science lecture room (Beale, 1890: 27). The original chemistry laboratory (Figure 3.6) dates from this same time (CLC Archives). Beale noted that in 1884: '… we fitted up more completely the Chemical Laboratory' (Beale, 1891: 11).

Figure 3.6: The 1876 Chemistry Laboratory at CLC

Source: Archives, CLC

In 1876 the examination was on heat and chemistry, and it was administered by P.T. Main, fellow and lecturer of St John's College, Cambridge. This must have been the first year the chemistry laboratory was used. In the examiner's report, it is commented that the chemistry theory part of the exam was not well done 'in the practical part, however, the result was satisfactory as shewing that many have acquired a habit of using chemical reagents intelligently, so as to draw correct positive and negative conclusions' (Anon., 1876: 17). He contended that the cutting back of theory to accommodate the practical had led to a deterioration of knowledge of chemical facts and principles.

A new wing of the building was proposed in 1903. It was designed by Millicent Taylor, the long-time chemistry teacher at the CLC (see below) and construction was completed in 1905. Taylor's report in *School World* (Taylor, 1905) on the new laboratories described in detail the dedicated chemistry lecture room with a fully equipped demonstration bench, an elementary chemistry laboratory, an advanced chemistry laboratory, a chemical preparation room, two storerooms for chemicals and equipment, and a chemistry study room. The chemical lecture room could hold up to 40 students and it was provided with a 15 ft-long lecture demonstration table, complete with sinks and fume hoods. The facilities were meticulously designed. The report also notes that, for the elementary chemical laboratory:

> On each bench are two light draught hoods ... Each bench has a separate gas and water control ... At one end of the laboratory is a slab of Yorkshire flag; for use with furnaces; at the other, draught cupboards and a hot closet. The floor of the hot closet consists of steel coated with copper, and is found most convenient for evaporations.
>
> (ibid.: 222)

The advanced chemistry laboratory (Figure 3.7) was similarly equipped but much more exact measurement was expected. To enable such high precision, extraordinary measures were taken to ensure the weighings had the utmost accuracy by isolating them from any air currents or vibrations:

> Two doors enclosing a well ventilated space lead from the laboratory to the balance room, which is far removed from the roadway as well as from all internal traffic. The floor consists of concrete covered with wooden blocks, and the slate balance slab is supported on pillars built up from the concrete foundations.
>
> (ibid.: 223)

Figure 3.7: The 1905 Advanced Chemical Laboratory at CLC

Source: Archives, CLC

Chemistry teachers

According to the prospectuses held in the CLC Archives, from 1867 until 1883, T. Wright, MD, FRS, FGS, was the teacher of physical geography and science. In addition, from 1868 until 1871, T. Bloxham, lecturer in chemistry and natural philosophy at the (boys') Cheltenham College taught natural philosophy at the CLC. An early student, Evelyn du Pré, recalled that it was Mr Bloxham who taught her chemistry (du Pré, 1910: 3).

The first woman science teacher at the CLC was Mary Watson (Vaughan, 1961: 1). Watson, born in 1856, attended St John's Wood High School (subsequently renamed South Hampstead High School), a GPDSC school. She entered Somerville College, Oxford, in 1879, obtaining honours in geology in 1882 and honours in chemistry in 1883. Watson accepted a position as science mistress at CLC in 1883 where she taught until her marriage to John Style in 1886.

Despite Watson's strong background in chemistry, it was Dr J. Norman Collie (Baly, 1943) who taught chemistry at CLC from 1883 until 1886. In a biography of Collie, his niece, Christine Mill, mentioned his departure from CLC: 'He was far from being a ladies' man and probably found that schoolgirls in bulk were rather more than he could stomach.' (Mill, 1987: 12).

It was Millicent Taylor who, in 1893, brought long-term stability to chemistry teaching at the school. Taylor was born in 1871 and attended CLC as a student between 1888 and 1893. It was from Cheltenham that she studied for an external BSc (London) degree which she completed in 1893.

At the same time, she was working towards a DSc (London), travelling to Mason College, Birmingham, at weekends for extra help (Anon., 1961). The London examiners considered she had not had time to cover all the material towards the DSc but they considered her work so good that she was offered a grant to study at a London college. However, her parents had moved to Cheltenham and she felt it was her duty to make a home with them. So instead she accepted an appointment to the staff at CLC. The following year, Taylor was made head of the CLC chemistry department, and then head of the science department in 1911 (Rayner-Canham and Rayner-Canham, 2008: 201–2).

Between 1898 and 1910 Taylor devoted most of her spare time to research work in organic and physical chemistry at University College, Bristol (later the University of Bristol) producing a range of papers in those fields. At weekends she would often cycle to and from the Bristol chemistry laboratories, an 80-mile round trip (Baker, 1962: 94). In addition, Taylor was one of the first group of women to gain admittance to the Chemical Society. Taylor received an MSc (Bristol) in 1910 and a DSc (Bristol) in 1911.

During the First World War Taylor was involved in production of ß-eucaine, a desperately needed local anaesthetic for operations on wounded soldiers. Then in 1917 she was appointed a research chemist at HM Factory, Oldbury. After the end of the war, in 1919, she returned to her post at CLC. In 1921, Taylor accepted a position as demonstrator in chemistry at the University of Bristol, being promoted to lecturer in 1923. Retirement in 1937 was not the end for her, as she was given the use of a small laboratory in an army hut on the grounds of the Bristol Chemical buildings. Taylor continued research until an accident in November 1960, her death following in December of the same year at age 89. During her lifetime, she was the author and co-author of a total of 19 publications, the last being published when she was 80.

The field club

The CLC field club was founded in 1889 (Innes, 1889). It had an archaeological section with 50 members; a botanical section with 30 members; a geological section with 16 members; and a zoological section with 20 members. Each of the sections was led by the appropriate woman teacher. The absence of a woman chemistry teacher until 1893 may account for the lack of a chemistry section, though none was established during Taylor's time either. Perhaps there was still the sentiment in places like Cheltenham that chemistry was certainly not for girls, whereas the discoveries of Mary Anning had made geology acceptable as a pursuit for women (Pierce, 2006).

Though chemistry had a comparatively low profile in the *Cheltenham Ladies' College Magazine*, there was one short story by a student, Pamela Meredith, about the life of a carbon atom (Meredith, 1927: 12). As we see in chapter 8, anthropomorphizing atoms and their chemistry was not uncommon in girls' school magazines.

Some rhymes at girls' schools referred to their chemistry teacher. At the CLC, a rhyme appearing in the *CLC Magazine*, referred to Millicent Taylor as 'A mistress skilled in Chemistry', and the opening lines are given below:

Ode to the Bottle of Phosphorous [sic]
Long ago in C.L.C.
    A mistress skilled in Chemistry,
Convinced that useful it would be,
    Bought a bottle of phosphorus.
Now it seems we're not so keen,
For see, this phosphorus has been
Stored here since grandma was sixteen –
    This ancient bottle of phosphorus.
I ask with what experiment
This mistress did herself content;
On Chemistry she was so bent
    She bought this bottle of phosphorus.
    ...

        (Moore, 1940: 12)

For the independent girls' day schools, the NLCS and the facilities of the NLCS chemistry lab was to be the benchmark by which the later schools were judged, while the independent girls' boarding schools looked to a different model, that of Cheltenham Ladies' College, for their evolution. Whichever template an independent girls' school followed, in order to replicate the original, either NLCS or CLC, chemistry and the need for chemistry laboratories was part of the requirement.

Chapter 4

Chemistry as a girls' subject

> ... *a certain number of girls have a real passion for science, are*
> *devoted to it, and often do very well later in college.*
>
> (Burstall, 1903: 8)

From the very beginning the teaching of science at the academic girls' schools was seen as an imperative in order to meet or exceed the teaching quality at boys' schools. Here we trace the discussions on chemistry for girls; on the heuristic method of teaching chemistry; and on the debate between teaching academic chemistry versus domestic chemistry.

Science for girls

One of the early strong proponents of a science education for girls was Sara Burstall, headmistress of the Manchester High School for Girls (see chapter 1). In a lengthy argument in *School World*, she began by noting that many, even 'enlightened', parents could not see the point of a science education for their daughters unless the girl was specifically interested in a career in science or medicine. Burstall, though being opposed to specialization of any kind, argued that: 'The value of scientific method, of verification and accuracy in observation, is in itself a corrective to the schoolgirl's fatal facility in learning up facts from a text-book, or mechanically reflecting the phrases and ideas of the teacher' (Burstall, 1903: 8). She believed that a laboratory experience played a particularly important role: '... it is all the more desirable that they should be obliged to get the training laboratory work gives – a peculiar and unique training, such as can be imparted in no other way' (ibid.: 8).

Burstall continued by pointing out the practical applications of science for the future woman homemaker:

> It is a very short-sighted and incomplete view which would consider general elementary science as useless in her education. All the various branches of domestic economy depend on the laws of physics, mechanics, and chemistry, from the frying of fish and the washing of flannels to sanitation and the care of children.
>
> (ibid.: 8)

Many of the headmistresses were proud of their school's success in science, even if they themselves lacked much science knowledge. A good example was Iris Brooks, headmistress at Malvern Girls' College. Grace Phillips, a former student, mentioned the importance of science at the college in her biography of Brooks:

> Miss Brooks's knowledge of Science almost equalled the vacuity of her information about Physiology or Mathematics. Nevertheless, she admired wholeheartedly the work of the Science Staff. She also rejoiced with them when Malvern Girls' College acquired such a reputation for achievement in scientific circles, that it became the first Girls' School to be awarded £10,000 from the Industrial Fund for the Advancement of Scientific Education in Schools.
>
> (Phillips, 1980: 216–17)

Chemistry courses for girls' schools

If chemistry was to be a focus of the independent girls' schools, it had to be at least as rigorous as that at the boys' schools. Thus defining content was a topic of specific concern.

Towards this goal, in 1884, for the schools of the Girls' Public Day School Company (GPDSC), the GPDSC Council distributed a list of which chemistry books were to be purchased for each school's library (Anon., 1884: 16). The titles of these books, together with their authors, are given in online Appendix 1. It is of note that two of the texts relate to practical laboratory work in chemistry, for as we see in chapter 6, the laboratory played a central role in the chemistry courses.

The GPDSC syllabi of 1896 and 1902

To ensure a consistency of excellence in the teaching of chemistry in all GPDSC schools, a *Conference on the Teaching of Science with Especial Reference to Chemistry* (Anon., 1896) was organized by the Council at Notting Hill High School in 1896. Chemistry teachers from all GPDSC schools were required to attend. Professor Henry Armstrong (see below) was one of the presenters, while the other was Ida Freund (see chapter 5), the leading woman chemistry educator from Newnham College, Cambridge (Palmer, 2012). Following from the conference, a detailed syllabus for the teaching of chemistry, including laboratory work, was produced by W.W. Fisher, Aldrichian Demonstrator of Chemistry at the University of Oxford

(Fisher, 1896). The complete listing of topics to be covered (see online Appendix 4) was circulated to all GPDSC schools.

Then in 1900, specifically for science teachers at GPDSC schools, a *Conference on the Teaching of Science* was held at Kensington High School (Anon., 1900). The major presentation was by Wyndham Dunstan, professor of chemistry at the Royal Pharmaceutical Society and previously lecturer in chemistry at the University of Oxford. He spoke on 'The Teaching of Science in High Schools for Girls' and following from his discourse, he was asked to devise a specific syllabus. His proposal was published in 1902 and reprinted in 1912 (Anon., 1902a). The course content is given in online Appendix 5.

Figure 4.1: A chemistry class at Portsmouth High School, 1908

Source: Archives, Portsmouth HS

The NLCS syllabi of 1911

In 1911 a book written by members of the Association of Head Mistresses was published entitled: *Public Schools for Girls: A series of papers on their history, aims, and schemes of study* (Burstall and Douglas, 1911). One of the chapters, written by Sophie Bryant, Buss's successor as headmistress of NLCS, discussed the teaching of natural science. Bryant noted that she was primarily addressing the senior-level specialized courses in science taken over two or three years. She also remarked that schools often had a further course for girls specializing in science.

Assisted by many science mistresses, in particular her colleague Miss Stern (see chapter 3), Bryant had collected information on the teaching of science – especially chemistry – from girls' schools across the country. She observed that, at the time, there were three distinct styles of teaching chemistry:

> Type I. – Based on the preparation of new substances from substances with which the learner is familiar, the examination of these substances after their preparation, and their action on one another.
>
> Type II. – Begins with the solution of a series of simple problems (suggested for the most part by the familiar facts of burning), which problems and their solutions are developed as far as possible from suggestions made by the learners.
>
> Type III. – The aim of which is the acquisition of knowledge leading up to the chemistry of domestic life, and the psychological spring of which is the learner's practical interest.
>
> (Bryant, 1911: 141)

Bryant considered that the first two types were theoretical-focused while the third appealed more to the 'practical mind'. The full course contents are given in online Appendix 6.

The heuristic method of teaching chemistry

Many – if not most – of the women chemistry teachers in the 1890s and 1900s became enamoured with the heuristic method of teaching chemistry, that is, using a discovery laboratory-based approach.

Origin of the heuristic method

It was the British chemist Henry Edward Armstrong (1848–1937) who first proposed teaching chemistry through discovery laboratory work. His first academic appointment had been as lecturer in chemistry at St Bartholomew's Hospital, London, in 1870 (van Praagh, 1973: 3). There, he was dismayed by the passivity of students. They simply wanted to memorize facts and definitions and whatever else was needed to pass the course. At the hospital he had to teach to the defined syllabus, but upon moving to a position at the London Institution at the end of 1870, he had much more freedom. It was there that he encouraged students to explore chemistry through experiments in the Institution's laboratory (Brock, 1973: 12).

Over the next decade Armstrong developed his ideas, testing them out on a boys' middle school in 1880. He named his teaching approach the heuristic method, 'heuristic' coming from the Greek 'find' or 'discover' and his personal definition of the term was: '… methods which involve our placing the students as far as possible in the attitude of the discoverer – methods which involve their *finding out* instead of being merely told about things' (Armstrong, 1903: 236).

To illustrate the use of the heuristic method, Armstrong described a study of the process of rusting. It is of note that he gives the example of the report by a 'lady teacher':

> Young children are delighted to be so regarded, to be told that they are to act as a band of young detectives. For example, in studying the rusting of iron, they at once fall in with the idea that a crime, as it were, is committed when the valuable strong iron is changed into useless, brittle rust; with the greatest interest they set about finding out whether it is a case of murder or suicide, as it were – whether something outside the iron is concerned in the change or whether it changes of its own accord.
>
> A lady teacher who had thus presented the case to a class of young girls told me recently that she had been greatly amused and pleased to hear one of the girls, who was sitting at the balance weighing some iron that had been allowed to rust, suddenly and excitedly cry out, '*Murder!*' This is the very attitude we desire to engender; we wish to create lively interest in the work and to encourage it to come to expression as often, as emphatically, as freely as possible.
>
> It is of no use for the teacher merely to follow an imaginary research path: the object must ever be to train children to work out problems themselves and to acquire the utmost facility in doing so … but do not let us spoil them by telling them definitely in advance what to look for and how to look for it: such action is simply criminal.
>
> (ibid.: 253–4)

The heuristic method at independent girls' schools

It was to be the British independent girls' schools who embraced heuristic teaching of chemistry with enthusiasm. Ironically, Armstrong was generally antagonistic to women in science. For example, he was the single greatest impediment to the admission of women chemists to the Chemical Society at a time when the large majority of British male chemists were supportive of the women's cause (Rayner-Canham and Rayner-Canham, 2003). Armstrong did not hide his very strong opinions on the topic of girls and women. In one comment, he stated: 'Those who have taught women students are one and all in agreement that, although close workers and most faithful and accurate observers, yet, with the rarest exceptions, they are incapable of doing independent original work' (Brock, 1973: 42). His opinion of women teachers was even more critical, contending it was essential for them to be married, because the single women teachers were 'the sexless creatures who too often engage in the vain task of training our daughters ... [they] are a real danger to society' (ibid.: 43).

Yet, as we will see, this same Armstrong was delighted to present his heuristic approach at meetings of girls' school science teachers. We can perhaps comprehend this apparent contradiction from a statement that he made to the educational science section of the British Association for the Advancement of Science in 1902:

> Experimental teaching is of even greater value to them [girls] than to boys, as boys have more opportunities of doing work which is akin to it in the world. The work done by girls should of course bear directly on their domestic occupations.
>
> (Armstrong, 1903: 91)

Figure 4.2: The chemistry laboratory at the Central Foundation Girls' School, undated

Source: Archives, CFGS

Armstrong particularly lauded the contribution of two of his former women research students who had become high school chemistry teachers: Grace

Heath at NLCS (see chapter 3); and Edna Walter at the Central Foundation Girls' School (see online Appendix 2). Armstrong visited the Central Foundation School in 1901 as was noted in the school magazine: 'Early in May we were honoured by a visit from ... Professor Armstrong, who examined the equipment of the chemistry rooms' (Anon., 1901: 11).

Heath's letter, to the journal *Nature*, gained a wide readership:

> By this new [heuristic] method the pupils themselves are put into the position of discoverers, they know why they are at work, what it is they want to discover, and as one experiment after another adds a new link to the chain of evidence which is solving their problem, their interest grows so rapidly, that I have seen at a demonstration lesson a whole class rise to their feet with excitement when the final touch was being put to the problem which it had taken them three or four lessons to solve.
>
> (Heath, 1892: 540)

The chemistry teachers at NLCS were to be among the most avid fans of the heuristic method. As described in chapter 3, Heath's successor as chemistry teacher at NLCS was Edith Aitken. Aitken was invited to address the annual meeting of the Association of Assistant Mistresses on the topic in 1898 (Aitken, 1898). She described how the heuristic method was introduced to 12-year-old girls at NLCS:

> We begin with such substances as sand and clay. Each child has her blue pinafore and her hair tied up out of the way of the gas flames. She has her stuff and does what she likes with it, subject, of course, to criticism and advice. She tries to dissolve, she boils it, bakes it, tries the action on it of acids, tries to crystallise it, examines it with a microscope, etc. ... If any one feels inclined to scoff at 'discoveries' made by little girls of twelve, I would beg them to try such a substance as gypsum with a similar class. I was myself quite surprised at the amount children managed to find out about the relation of Plaster of Paris to Gypsum and its uses.
>
> (ibid.: 8–9)

Aitken, in turn, was replaced by another supporter of the heuristic method, Rose Stern (see chapter 3). Stern, together with her friend Alice Maude Hughes, science mistress at Eltham Hill Secondary School for Girls, wrote a laboratory manual along heuristic principles: *A Method of Teaching Chemistry in Schools*. In the preface, they lay forth their principles:

> ... it is intended that every experiment should be suggested and
> carried out by the pupils, the part of the teacher being only to
> guide and supervise. At the same time the teacher must reserve
> the right of selecting the experiment to be done by the class
> when several have been suggested, and, in this way, preventing
> time being wasted in trying experiments which would be of little
> value to the children and which would break the sequence of
> their work.
>
> (Hughes and Stern, 1906: v)

The use of the heuristic method at British independent girls' schools became accepted practice. For example, in a history of Bedford High School, it is commented: '... when Professor Armstrong and his Heuristic Method ... had caused a good deal of fluttering in the scientific dove-cote, it became absolutely necessary to make some provision for individual practical work' (Westaway, 1934: 74).

It is not surprising that Blackheath High School was another of the many girls' schools to adopt the heuristic method as Armstrong had sent his daughters there. The compilers of the history of the school, Mary Malim and Henrietta Escreet, described the use of the heuristic method to promote the scientific method. In a footnote they cite a comment by the science teacher Sophie Smith: 'It [the heuristic method] appealed to many Science teachers as providing a very sound basis for elementary work in Science, although not altogether suitable for the more specialized work of older students' (Malim and Escreet, 1927: 46).

Opposition to Armstrong's heuristic method

However, there were voices of dissent. The famous British chemist Sir William Ramsay was one of those who vociferously espoused the lecture method and expressed a dismissive hostility towards the heuristic method (Anon., 1902b: 328).

More important to the teaching of chemistry at girls' schools, Freund did not agree with Armstrong's approach. She wrote in her book, *The Experimental Basis of Chemistry: Suggestions for a series of experiments illustrative of the fundamental principles of chemistry*: '[This] would have us believe that in the course of some couple of hours' work the average pupil can definitely correlate an observed effect with its cause, can *discover* the nature of a chemical relationship, or can prove a law' (Freund *et al.*, 1904: 6).

Freund preferred the approach of Wilhelm Ostwald. In her book she describes the Ostwald educational philosophy:

... the main facts of inorganic and physical chemistry are dealt with in the form of dialogues between a teacher and a pupil. The method is heuristic in the truest and best sense, but there is no make-believe, no pretence about what the pupil really accomplishes himself and what is done for him.

(ibid.: 7)

Decline of the heuristic method

To use guided inquiry required considerable skill on the part of the teacher: to guide towards the goal, not to leave the students aimless. However, as Armstrong's heuristic devotees retired, the new generation lacked the training in the proper context and application of the method. As a result, by the 1920s, the heuristic approach was in decline.

The cause of the decline was explained by Dorothy Mabel Turner in her book, *History of Science Teaching in England*:

Unfortunately the disciples of Armstrong went too far. They regarded practical work in the school laboratory as an end in itself ... They were afraid to tell their pupils anything, and the unfortunate young investigators often gained nothing from their work in the laboratory but a marked distaste for the subject. The over emphasis on method and the ignoring of the importance of the content has done much to bring heuristic teaching into disrepute.

(Turner, 1927: 145)

In fact many later British science educators believed that the lecture method, interspersed with demonstrations, was a far more effective use of the teaching time and resulted in better learning outcomes (Connell, 1971).

Academic chemistry or domestic chemistry?

In the later nineteenth century domestic studies were seen as a key component of girls' education. Initially, it was domestic economy – cooking, laundry work, and so on. Two subsequent reports, the Interim Report on Housecraft in Girls' Secondary Schools in 1911 and the Consultative Committee on Practical Work in Secondary Schools in 1913, both recommended that, in the new 'scientific age', the teaching of domestic subjects should have a strong foundation in science and become domestic or household science (Dyhouse, 1977). Henry Armstrong had a strong opinion, arguing that both boys and girls would benefit from the scientific study of domestic issues (Bayliss, 1983).

At the core of domestic science was chemistry – especially the chemistry of foodstuffs and household cleaners. Tertiary-level institutions and departments were founded that focused on such an education. These establishments existed across the country, for example in London they included the Training College of Domestic Subjects, Berridge House, Hampstead; the School of Home Training in Domestic Science, South-Western Polytechnic, London; the School of Domestic Economy, Battersea Polytechnic Institute; and the Home Science and Economics Department of the Women's Department of King's College, University of London (Rayner-Canham and Rayner-Canham, 2011).

Independent girls' schools had seen their main mission as sending as many graduates as possible to university studies. However, in the early twentieth century, even in those schools debate began: should girls study domestic science for their role as homemaker, or academic science – particularly chemistry – to provide them with career options (Manthorpe, 1986; Spencer, 2000)?

We must keep in mind that this period was very different from our own. The expectation of the girls from independent schools was that, if they did marry, the day-to-day work would be performed by maids. Thus there seemed little relevance for them to study household science. However, the biochemist Ida Smedley Maclean (see online Appendix 3) argued that formal knowledge of some domestic science was not only important for the staff but also for the lady of the house:

> The scarcity of domestic servants makes it possible for a girl with practically no training whatever to obtain a place as a cook or housemaid … A comparatively short course of training in the household arts … should prove of great help, and should enable the woman whose professional occupation consists in the management of her house and children to give some adequate training to her maids if she be not so fortunate as to secure well trained ones.
>
> (Maclean, 1914: 338–9)

Some independent girls' schools had been offering homecraft courses for some years, in addition to academic science courses. The transformation of domestic studies to a more scientific focus was noted for the Domestic Science programme at Cheltenham Ladies' College (CLC):

> The course of instruction in Domestic Science has been altered in order that it may be placed on a more scientific basis, so that

a student may become so thoroughly acquainted with the nature and properties of the materials she has to deal with that she is able to work from this knowledge and not merely empirically.

The syllabus consists of specially adapted courses in Physics, Chemistry, Physiology and Hygiene, while special Chemical work relating to food-stuffs is done in a separate laboratory in the kitchen.

(Anon., 1908: 261)

Figure 4.3: Students performing domestic chemistry experiments at Mary Datchelor School

Source: Whyte, 1901–2: 269

The influence of Arthur Smithells

It was the push to make domestic science into an academic discipline that aroused debate. A leading proponent of the teaching of domestic science to girls was Arthur Smithells, professor of chemistry at the University of Leeds. He had been active in promoting the science education of girls as a result of briefly teaching at Manchester High School for Girls while he was at Owens College (later Manchester University):

> Part of his last session (1882) at the Owens College was spent in conducting a course of lectures and practical work on chemistry at the neighbouring Girls' High School where he gained his first experience of teaching, an experience which may have directed his attention to the gaps in a girl's education, and implanted the germ of his present schemes for the scientific training of women.
>
> ('J.B.C.', 1910: 37)

Smithells was part of the 'Science for All' movement that was concerned with the low level of scientific awareness among the general population (Flintham, 1977). Members of the movement contended that humanization of science was the answer, in which scientific principles were related to people's daily lives. Smithells saw domestic science as a means of bringing an applied aspect that would, in particular, be appropriate in the education of girls (Flintham, 1975).

In his contribution to the discussion, Smithells was very circumspect, simply contending that if teachers were not constrained by rigid curricula, then they would teach science in a broader context (Smithells, 1912: 460).

Smithells had the ardent support of Miss M. Wood, science teacher at Leeds Girls' High School. The *Journal of Education* carried a report of the 1909 meeting of the Association of Assistant Mistresses in Public Secondary Schools. In it, they described Miss Wood's attempt to:

> ... carry out the ideas of Prof. Smithells, by connecting domestic science with the science teaching of the school and applying the scientific method to common things. She gave the details of an ideal scheme of theoretical and practical work in a 'kitchen laboratory', which should extend from eleven to sixteen, and by which the girls were to acquire both the training in scientific method and the most important points of domestic hygiene, and be ready to take up preparation for University work in some branch of science.
>
> (Anon., 1909: 149)

Wood subsequently described the entire five-year programme that she had implemented for girls from the age of 11 to 15. She attributed the success of the programme to the guidance of Prof. Smithells and the construction of a 'kitchen laboratory' at Leeds Girls' High School (Wood, 1909: 140–3).

Proponents of domestic science

It was in the years from 1909 to 1912 that the discourse became highly public and polarized. There were two educational publications within which debate raged: the *Journal of Education* and *School World*. Here we highlight some of the different viewpoints.

First, there were some strong proponents for domestic science, in particular for a separate stream for the non-academic student. This view was espoused by Lucy Hall and Ida Grünbaum, science lecturers at Avery Hill [Teachers'] Training College, Eltham, who subsequently co-authored *The Chemistry of Housecraft: A primer of practical domestic science* (Hall and Grünbaum, 1912). They stated that the chemistry syllabus commonly taught in schools, with its emphasis on calculations, was not appropriate for some girls who would otherwise gain much benefit from a study of chemistry:

> One type of girl, always to be found in a school, is the sensible, practical person not endowed with too many brains for books. To such a girl a chemistry lesson would be a delight, and would afford opportunity for developing her practical powers, while at the same time, she would be quite able to appreciate the reasoning. She (at thirteen or fourteen) would work out the experiments on the composition of chalk and washing soda with great interest and common sense, whereas she would be completely fogged by the calculation of the latent heat of steam.
>
> (Hall and Grünbaum, 1910: 105)

Florence Storr, science mistress at the Central Foundation Girls' School from 1897 until 1915, insisted that a strong background in chemistry and physics in middle school was an essential prerequisite for any training in household science (Storr, 1910). She provided details of a course to be taken at age 15 in chemistry, hygiene, and cookery that needed to be taught by a science teacher to ensure everything was founded upon scientific principles. She noted: 'So that the girls' secondary school of the future will not only equip the future university student and teacher, but will consider that preparation for home life – the life-work of the majority of women – is a fundamental part of its work' (ibid.: 407).

Lilian Faithfull, who succeeded Beale as principal of CLC, concurred:

> I am of opinion that science classes, i.e., physics and chemistry, should be taken in a school from the age of twelve and thirteen, independently of the application of science to domestic work, and

that a good foundation of scientific method should be laid first of all, so that this work may be equally useful whether the girl at a later state of her career elects to follow a university course and to take the degree of Bachelor of Science, or to enter the home science department and learn housewifery and the chemistry of cookery.

(Faithfull, 1912: 452)

Proponents of academic science

Having fought hard for getting girls an academic education equal to that of boys, most women science teachers opposed any shift of science teaching towards a domestic context. They saw it as dangerously limiting girls' aspirations and opportunities and turning back the clock to the expectations of domesticity. Leading the rebuttal, Ida Freund savagely attacked the teaching of domestic science as a 'pseudo-science':

> The general public, educational authorities, and, alas! not a few headmistresses have been dazzled by promises of a combination between the educational and the utilitarian, of housecraft and scientific method learnt at one and the same time, of the advent of a new era in which every woman will know all that is known (and more) about food values and the chemistry of cleaning. On the other hand, a large number of experienced women science-teachers, among whom not a few can claim first-hand experience of household work and household management, whilst as keen as anyone can be about raising the status of domestic work and spreading a knowledge of practical hygiene, see in the almost unprecedented popularity of this new brand of pseudo-science a real and serious danger to girls' education.

(Freund, 1912: 453)

Edith Lees, senior science mistress at Clapham High School, reiterated Freund's argument:

> Whilst being strongly in sympathy with the movement to teach girls domestic matters, I believe they should be taught as arts, at a time when the arts can be practised. In the schemes and syllabuses I have seen of 'Science of Home Life' or 'Domestic Science' I have been unable to recognize either good art or good science.

(Lees, 1912: 456)

Hilda Jane Hartle (see chapter 5), at that time lecturer in chemistry at Homerton College, Cambridge, argued that little was then known about the chemistry of household life:

> The science of cookery and of laundry work is yet in its infancy. No literature of the subject exists. Not even the most brilliant organic chemist can be said to 'know' the chemistry of foods, still less can such a subject be within the grasp of students in training.
>
> (Hartle, 1911: 849–50)

Six months later Hartle wrote a scathing attack on a textbook, *Experimental Domestic Science*, written by R. Henry Jones, head of the chemistry department, Harris Institute, Preston, in which Jones had claimed that 'science can be directly and adequately taught in the kitchen, and that a previous training in elementary science is not indispensable' (Hartle, 1912: 465). Hartle not only attacked the book in principle, but also pointed out the many errors in chemistry that it contained.

Another science book of the time aimed at girls was the *First Book of Experimental Science for Girls*. The author, Jessie White, had completed the Natural Science Tripos at Newnham College (Anon., 1904: 172), where she had been one of Freund's students. At the time of her retirement in 1905, White had been headmistress of the High School for Girls at Highbury Hill House. The book used examples from the kitchen, laundry, and other parts of the house, to illustrate scientific principles. There had been a positive review of the book in the *Journal of Education* in 1914 that prompted an irate response from Isabel C. Fortey (see online Appendix 2):

> The review in your August number of Mrs. White's "First Book of Experimental Science for Girls" begins with the words, "To those who have studied the problem, it has for long been obvious that girls require a course in science quite different from that which it has been customary to provide for boys." Such a statement will not be allowed to pass unchallenged because Miss Freund is dead.
>
> (Fortey, 1914: 618)

Isabel Fortey, sister of chemist Emily C. Fortey (Rayner-Canham and Rayner-Canham, 2008: 203–4), then argued that a girl had to do the same science as her brother in order to compete with him at science studies in university, or if she wanted to become a doctor.

The crucial importance of inculcating the 'scientific method' into young girls was emphasized in a joint submission by F. Gadesden, headmistress, and S. Frood, science mistress of Blackheath High School:

> The chief aim in teaching science in schools should be to inculcate the scientific method. Principles should be taught rather than details ... We entirely agree with the opinion that domestic "operations" should be taught as home arts, and that their scientific treatment is more suitable for postgraduate study.
>
> (Gadesden and Frood, 1912: 454)

The headmistress of Croydon High School, E. Marion Leahy, tried to find middle ground. Leahy made a plea that, though academic science was important, the applications of science should be discussed whenever possible:

> There can surely be only one road for beginners in elementary physics and chemistry, and we do not wish to alter the sound scheme of training which is doing so much to develop the intelligence of our girls. But we do ask that there shall be a very sympathetic attitude on the part of our science mistresses to the general purpose of our girls' education. We ask that, whenever possible, illustrations may be taken from everyday experience in the home, and the girls' attention constantly directed to experimental work in connection with such experience. We ask that it may be borne in mind that very few of our girls are going up to the university to study science, but that all of them are going to be women. Every woman needs to specialize as a home-maker, whether she is, in the future, married or single, working at a profession with a little home of her own, managing a large household of children and servants, presiding over some institution, or taking part in social and philanthropic work.
>
> (Leahy, 1912: 456)

In the end, with the antipathy of most of the women science teachers, domestic science – specifically domestic chemistry – ceased to be an acceptable alternative science subject for academic girls. As an illustration of the subject's demise, the domestic science department of Sutton High School (a GPDSC school), opened in 1911, then closed in 1916 (Spencer, 2000: 80). There had been discord over what 'domestic chemistry' entailed, and the examining universities never offered it as an academic subject (Davies, 1923). With the pressures for academic success at girls' schools, domestic chemistry was relegated to a low-status non-academic subject. The high school debate was over.

In this chapter we have mentioned the names of several women chemistry teachers. Where did they obtain their education? What was life like for them? This is the topic of the next chapter.

Chapter 5

The pioneering chemistry teachers

> An Impending Departure (to Miss Ready, Senior Science
> Mistress 1901–32)
> Despite the laws of Physics she
>   has taught us,
> These many years, with patience
>   rarely known,
> Immersed in our affections,
>   she displaces
> A volume so much greater than
>   our own.
>
> (Anon., 1931–2: 6)

To produce such enthusiastic chemistry students required dedicated and equally enthusiastic women chemistry teachers. These teachers were the first generations of women to go to university. Here we look at the available educational pathways to becoming a chemistry teacher and the challenges it entailed. Then, in online Appendix 2, we provide biographical accounts of some women chemistry teachers, who, like Miss Ready in the poem above, devoted their careers and lives to the cause of science teaching, especially chemistry.

The earliest chemistry teachers

In the earliest times of our account, there were no women trained in chemistry. Thus teaching was accomplished by men who gave of their time and effort, presumably from a belief in the rights of girls to learn science. In chapter 2, we showed that at Newington Academy for Girls it had been William Allen; at the Mount School, Edward Grubb; and at Polam Hall, Robert Sargent. At North London Collegiate School, it had been Robert Buss, and at Cheltenham Ladies' College, T. Wright and T. Bloxham, as mentioned in chapter 3.

At St Swithun's School, about the late 1880s and early 1890s, it was Revd H. Searle who provided chemistry and physics teaching to the students:

The Rev. H. Searle was asked to give us lessons in Physics and Chemistry. You to whom the possession of a well-equipped laboratory is commonplace can have no idea of the thrill of those early lessons given in the School Hall ... We had no apparatus and he had to carry everything, flasks, chemicals, heavy iron apparatus, by bicycle to and from the Training College, and yet from the first every lesson was fully illustrated by experiments.

(Finlay, 1934: 19)

Similarly, at Sheffield High School during the earliest years of the school, a male teacher was 'imported' to teach chemistry. In the *Sheffield High School Magazine*, one 'old girl' reminisced: 'Chemistry was taught by a gentleman from Chesterfield: his h.'s were apt to go astray, but he made the lessons interesting and told thrilling stories of duchesses who bleached their hair with peroxide ...' (Anon., 1928: 6).

Though briefly preceded by a woman chemistry teacher, Mary Adamson (see online Appendix 2), George Samuel Newth taught chemistry in about 1890 at Princess Helena College. Newth was a demonstrator and lecturer in chemistry at the Royal College of Science (later Imperial College) from 1871 to 1909. He wrote several chemistry books including *Chemical Lecture Experiments* (Wikipedia contributors, 2015a). An issue of the *Princess Helena College Magazine* of 1890 reported that: 'Mr G.S. Newth [was] teaching chemistry' (Anon., 1890: 9). In a history of Princess Helena College, it is noted that Newth taught chemistry using 'numerous and beautiful experiments' (Clarke, 1991: 164).

An alternative solution, of sending the students elsewhere for chemistry, was used at some schools. A documented example is that of Leeds Girls' High School, founded in 1876. The first headmistress, Catherine Kennedy, was a strong proponent of chemistry as the key science for girls. However, she realized that there were few, if any, women teachers qualified to teach the subject (Jewell, 1976: 47–8). So she arranged for the students to take their chemistry classes and laboratory sessions at the Yorkshire College of Science (later the University of Leeds). The chemistry professor at the college was T. Edward Thorpe, a strong supporter of women in chemistry (Rayner-Canham and Rayner-Canham, 2008: 71). The girls excelled in chemistry, both that first year and in subsequent years. For example, in the obituary for a former student, Edith Little, in the *Leeds Girls' High School Magazine*, it was noted that:

... in 1880 she [Edith Little] gained the Prize and First Certificate for Chemistry of the Non-Metals and Practical Chemistry at the

> Yorkshire College, which the older pupils of the school used to
> attend at that time for instruction in chemistry ...
>
> (Kennedy, 1908: 5).

The arrangement continued only until 1883, even though the school laboratories were not completed until 1907. Most likely Thorpe was the only chemistry professor willing to teach girls, and by 1884 he had an increasing interest in field research. Then in 1885 he moved to a professorship at Imperial College, London.

It was a neighbouring boys' school that provided the solution for St Martin-in-the-Fields High School for Girls, then in Charing Cross Road, London. Twice a week the girls were walked to the chemistry laboratory of nearby Archbishop Tenison Boys' School. They left their school at 4 p.m. arriving at Tenison's after the boys had left, and returned to St Martin's at 5 p.m. From 1903 until 1916 the laboratory sessions were taught by the Tenison chemistry master, William Henry Ratcliffe (Staff Register, St Martin-in-the-Fields Archives). In her reminiscences Lydia Mentasti, who started at the school in 1897, recalled: 'Girls were chaperoned to Archbishop Tenison Boys' School for chemistry. There were usually no boys about, but it would have been against the rules to speak to them if there were' (Siddall, 1999: 40).

The Headmistress of St Martin-in-the-Fields, Miss C.C. Derrick, appointed in 1874, had been assistant mistress at North London Collegiate School under Buss, thus it is not surprising that she demanded academic excellence – including in chemistry – at her school (ibid.: 36). It was not until the school moved to the London suburb of Tulse Hill in 1928 that finally the girls had their own spacious and well-equipped chemistry laboratory (Cheeseman, 1928: 24–5).

Ackworth School, a Quaker school near Pontefract, West Yorkshire, had a boys' school and a girls' school side by side. Though the chemistry laboratory was built in the boys half, girls had a weekly chemistry lesson there with a male chemistry teacher by 1889. Elfrida Vipont notes that in 1893 'an old scholar described his astonishment at seeing "a whole bevy of girls invading the boys wing ... to join a combination chemistry class"' (Vipont, 1959: 128). It was in 1909 that a specialist in mathematics and science, Jane H. Williamson, was appointed to the girls' school (ibid.: 156).

Women's colleges producing chemistry teachers

We were able to find some information on the origins of many of the women chemistry teachers. Perhaps not surprisingly, nearly all of the women chemistry teachers came from independent girls' schools themselves.

School–university linkages

The staff registers of the respective schools always noted the school at which the teacher herself had obtained her education. The majority of the women chemistry teachers-to-be had attended one of the two Cambridge women's colleges: Newnham College or Girton College. To a lesser extent, they continued their studies at one of the two London women's colleges: Royal Holloway College or Bedford College. The reader may be surprised that only a very few attended any of the Oxford women's colleges. In fact Oxford was seen by young women as an arts-focused university in general (Howarth, 1987).

For some schools there was a strong link with a particular university. For example students often progressed from Blackheath High School, from James Allen's Girls' School, and from Cheltenham Ladies' College, to Girton College. Students completing their school education at King Edward VI High School for Girls usually went to Newnham College, while those from Wimbledon High School often entered Royal Holloway College. For the schools outside the south, the students tended to continue their education at the nearby university college, for example, many students from Sheffield High School went to University College, Sheffield (later Sheffield University) and those at Manchester High School for Girls to Owens College, Manchester (which was merged into Manchester University).

The Cambridge women's colleges

The opening of Girton (1873) and Newnham (1875) had been inspirational beyond the actual numbers of students who attended. This is illustrated by a comment of Winifred Sturge and Theodora Clark, who authored the history of The Mount School, York:

> No other public event [the opening of Girton] can have more influenced the Mount during our period. To Lydia Rouse [headmistress at The Mount School] it was of greatest moment. She set constantly before her schoolgirls the prospect of college life as a practicable ambition, and they strained to look through her window, standing not without effort on tiptoe.
>
> (Sturge and Clark, 1931: 108–9)

Initially, there were no chemical laboratory facilities at either Girton or Newnham College. During the 1870s Philip Main had generously organized early morning practical chemistry sessions for women students at his laboratory at St John's College (Gardner, 1921: 37). His classes ceased in 1879 with the completion of chemistry laboratories at Newnham and Girton. With their own facilities within the walls of the women's colleges, women chemistry students at Cambridge no longer needed chaperones to attend the laboratory sessions.

Newnham College

From 1887 Ida Freund 'reigned supreme' in Newnham's own chemistry laboratory (Wilson, 1988: 72). In chapter 4 we described the influence of Freund on the debate between academic chemistry and domestic chemistry. Born in 1863 in Austria, Freund studied at the State School and the State Training College for Teachers in Vienna. In 1881 she moved to Britain to join her uncle and guardian, the violinist, Ludwig Strauss ('E.W.', 1914).

Freund was sent to Girton for her university education (Hill and Dronsfield, 2004). Completing her undergraduate studies in 1886, Freund accepted a one-year lectureship at the Cambridge Training College for Women (see below). The next year she was offered a position as demonstrator at Newnham and promoted to lecturer in chemistry in 1890. In a letter to her mother, Catherine Holt, one of the students of 1889, mentioned Freund:

> I attended my first lecture yesterday; it was Chemistry; there were about 8 students from this college and three from Girton ... Afterwards we adjourned for a couple of hours to the laboratory here; Miss Freund is the presiding genius, a jolly, stout German, whose clothes are falling in rags off her back.
>
> (Holt, 1987a: 1)

In a subsequent letter, Holt expanded on her chemistry experiences in Freund's laboratory:

> ... we got arsenic and phosphorus fumes at the [chemistry] lecture yesterday morning and a frightful smell of ammonia at the Laboratory afterwards. It was perfectly disgusting. All the same Chemistry is great fun and I did some splendid experiments yesterday.
>
> (Holt, 1987b: 13)

Figure 5.1: Ida Freund and the Newnham Chemistry Laboratory
Source: Archives, Newnham College, Cambridge

Another chemistry student, Marguerite Ball, recalled that the standard attire of Newnham students was also worn in the chemistry laboratory: 'We wore almost a uniform of white blouse, and tweed coat and skirt ... When I wore my first, my clumsy spilling of acid in the lab. made large holes in the front which no darning would conceal ...' (Ball, 1988: 77–8).

Freund retired due to ill health in 1913. The year of her retirement the Newnham chemistry laboratory was closed, as women students were then given access to the main Cambridge chemistry laboratories. She died during a medical operation in 1914.

Girton College

During the period 1897–1906, Dorothy Blanche Louisa Marshall was the demonstrator in the Girton chemistry laboratory (Butler and McMorran, 1948a: 638). Marshall was born in London in 1868. She was educated at King Edward VI High School for Girls (KEVI), and then went to Bedford College in 1886. From there Marshall transferred to University College, London (UCL), graduating with a BSc in 1891. Staying at UCL as a post-graduate researcher until 1894, Marshall then accepted a one-year demonstratorship at Newnham. In 1896 she was appointed demonstrator in chemistry at Girton, and promoted to resident lecturer in chemistry in 1897.

Marshall left Girton in 1906 to become senior science lecturer at Avery Hill Training College, though science was not actually taught there until the 1930s (Shorney, 1989: 53). Soon after being promoted to acting principal, Marshall became ill and resigned. During the First World War, she was made a scientific research assistant at the National Physical Laboratory where she stayed for the remainder of her working life.

Marshall was followed at Girton by another former KEVI student, Mary Beatrice Thomas. Just as Freund had 'reigned supreme' at Newnham, so her close friend Thomas was a dominant figure in chemistry at Girton from 1902 until 1935. Thomas was born in Birmingham in 1873 (Rayner-Canham and Rayner-Canham, 2008: 230–1). She followed the well-trodden trail from KEVI to Newnham in 1894 where she studied chemistry and physiology, completing the degree requirements in 1898. After graduation she was a demonstrator in chemistry at Royal Holloway College.

From 1902 to 1906 Thomas was demonstrator in chemistry at Girton College under Marshall. Following Marshall's departure Thomas was appointed lecturer and director of studies in natural science at Girton, a post she held until 1935. A student, 'D.M.P.' described in the *Girton Review* how Thomas was a scientist and a feminist:

> … an eager devotee of the gospel of dedication to that search for truth and for scientific knowledge which, at the end of the Victorian age, had for many ardent spirits taken the place of orthodox religion. We found her unyielding in her demands that we should put academic tasks before all other interests, that we should maintain the tradition of the earnest pioneers, who never for one moment allowed themselves to forget their goal of proving and demonstrating that no distinction can be drawn between the intellectual powers of men and women, once they are offered similar educational opportunities.
>
> ('D.M.P.', 1954: 17)

Royal Holloway College

In the early years the chemistry department at Royal Holloway College (RHC) was dominated by women chemists. The department of chemistry had been founded in 1891 by M.W. Robertson, but it was Elizabeth Eleanor Field who was head of the department for two decades (Anon., 1932a). Field had been educated at Newnham College, graduating in 1887 (Fortey, 1933). From 1889 to 1890 she was an assistant demonstrator in chemistry at Newnham. After leaving Cambridge she held a post as assistant mistress for two years (1893–5) at Liverpool College for Girls and then became lecturer and head of chemistry at RHC. Women heads at the RHC were titled 'senior staff lecturer' while the male faculty of the same rank were titled 'professors'.

After her retirement in 1913 she moved to Brno, Czechoslovakia, where she died in 1932.

According to an obituary published in the *College Letter of the Royal Holloway College Association*, Mary Boyle was another stalwart of the chemistry department at RHC (Anon., 1944). Born in 1874 she entered the Royal College of Science in 1898. Boyle took the full three-year course for chemistry before transferring to RHC in 1901, where she gained her BSc after one year of study. She remained at RHC, firstly as a demonstrator, then assistant lecturer in 1906, subsequently being promoted to staff lecturer. Boyle retired in 1933, moving to Leeds where she died in 1944.

Bedford College

It was the arrival of Holland Crompton in 1888 as head that really marked the founding of the department of chemistry at Bedford College (Harris, 2001: 81). One of Crompton's first tasks was the design of the new laboratory. An article in *The Girl's Realm Annual* reported: 'Further up the stairs you come across the chemistry laboratory, where Mr Crompton is superintending a large class. Each girl wears a smock-pinafore of coloured linen' (Rawson, 1899: 927).

In chapter 6 we focus on the school chemistry laboratories and in several of the student recollections, fires and explosions are mentioned. Likewise in chapter 8, where we cite some chemistry poetry and prose written by students, allusion was sometimes made to accidents in the chemistry laboratory. From the way in which such incidents are described, it is clear that chemical accidents were regarded as an intrinsic part of doing school laboratory chemistry work. This somewhat blasé attitude extended to the university laboratory environment, as the following commentary in an 1898 issue of the *Bedford College Magazine* shows:

> A certain amount of excitement was caused one afternoon by the fact that one student was suddenly seen to be in flames. However, she lay down quite calmly, and was immediately knelt upon by her nearest neighbours, so that all danger was over before most people knew what had happened and they only caught a glimpse of her as she lay 'smiling and smouldering' on the floor. We consider the behaviour of those concerned a credit to the College and to the cause of women's education.
>
> ('H.A.B.', 1898: 16)

Trinity College and the 'Steamboat Ladies'

Women were barred from formal university degree status at Oxford until 1920 (Rogers, 1938) and at Cambridge until 1948 (McWilliams-Tullberg, 1975). However, for a brief period from 1904 to 1907, women who had completed the Tripos requirements at one of the Cambridge or Oxford (Oxbridge) women's colleges were eligible to be granted an MA degree from Trinity College, Dublin (Parkes, 2004: 87–8). Of the chemistry teachers whose names were listed in staff registers that we studied, a high proportion of Oxbridge graduates of the time had availed themselves of this opportunity to obtain formal degree status.

For many years Trinity had enjoyed close relationships with Oxford and Cambridge, in particular the reciprocal recognition of examinations. In the early years Trinity did not admit women; so many Irish women went to women's Oxbridge colleges for their education. Thus it seemed unfair to many that Irish women who had successfully completed the degree requirements at Oxford or Cambridge should not have any formal degree title. To right this wrong the new provost of Trinity College, Anthony Traill, a strong supporter of women's higher education, proposed that Oxbridge women graduates be granted Trinity degrees. In June 1904 the Senate approved such a move with a termination of December 1907, by which time Trinity would be producing its first class of women graduates.

Traill, and the Trinity Senate, had assumed that, among the women Oxbridge graduates, only those of Irish ancestry would take advantage of the opportunity, but events proved otherwise. Word spread rapidly. The women's colleges at both Oxford and Cambridge encouraged all of their current and earlier graduates to take advantage of the situation. The Clothmakers' Company, which had provided scholarships for many of the women students at Girton and Newnham, Cambridge, and Somerville, Oxford, announced it would pay the commencement (graduation ceremony) fees of any woman who wished to obtain her degree letters from Trinity.

During the nearly four years this dispensation was in effect approximately 700 Oxbridge women availed themselves of the opportunity to purchase a formal degree. The whole expedition could be accomplished in as little as 34 hours using the fast mailboats. Women graduates often arranged to travel as a group of friends from the same graduation year in a college, or from the same school in which they were currently teaching. Caroline Coignou, chemistry teacher at Manchester High School for Girls (see online Appendix 2), had completed the Tripos requirements at Newnham College and she reported about her brief sojourn in Dublin in the *Magazine of the Manchester High School*:

On Saturday April 29ᵗʰ, [1905] a good number (87) of women from Oxford and Cambridge – mostly from the latter University – assembled in "No. 6" [an address in Dublin] to don their gowns and hoods. It was a merry gathering; old friends who had not met since College days greeted each other.

('C.C', 1905: 50–1)

Teachers' training colleges

It was the accepted opinion among men planning a teaching career that, like their own teachers, an academic background was all that was needed to become successful (Bottrall, 1985: 4–6). Some of the women teachers had a different perspective and wished to gain a knowledge base in the theory of teaching. As a result, in the late 1800s and the early 1900s, a large number of women-dominated teacher training colleges and departments were instituted across the country (Edwards, 2001). Among the earliest were the Maria Grey Training College in London founded in 1878 (Lilley, 1981); and the Cambridge Training College for Women Teachers (Bottrall, 1985) and Edge Hill College, Liverpool (Montgomery, 1997), both founded in 1885.

None of the women chemistry teachers for whom we have biographical information attended Edge Hill College. The offering of chemistry received only a passing mention in the history of the college but there was an interesting comment on a science society that existed about 1910: 'The Science Society was heavily academic: it was founded by the Third Year Science Students (i.e. the university students) and some of the "Normal Class"' (ibid.: 33). Average attendance was around 20, 'as only those really interested in science are desired' (ibid.: 33).

For validation most of the teachers' colleges relied on the examinations offered by the Cambridge Teacher Training Syndicate (CTTS). Thus a Cambridge Teaching Diploma or Certificate rarely indicated that the woman teacher had actually attended Cambridge. Cambridge University charged a considerable fee for external students taking the examination, but the Diploma/Certificate from the CTTS provided the graduates of other teachers' colleges with national credibility in their teaching qualification.

The aim of combining a depth of academic knowledge with the theory and practice of teaching was admirable. About 25 per cent of the 128 women chemistry teachers at independent girls' schools for whom we have biographical information added a teaching qualification to their academic degree. Irrespective of where they studied teaching methodology and practice, almost all completed the Diploma or Certificate of the CTTS.

Nevertheless, the majority of women chemistry teachers went directly from a university degree to a teaching position. In a biography of Buss, Sophie Bryant remarked: '... women educated at the Universities persisted in neglecting professional training. Either they despised it, or they could not afford it, or they did not like it, and could get entrance to the schools without it' (Ridley, 1896: 283).

Maria Grey Training College

Maria Grey had played an influential role in the formation of the Girls' Public Day School Company (GPDSC). She realized that to sustain strongly academic girls' schools there was a need for highly trained women teachers (Ellsworth, 1979). To that end Grey was one of the leading figures involved in the founding of the Teachers' Training and Registration Society College that opened in Bishopsgate, London, in 1878. The college prospered after the arrival of the new principal, Agnes Ward, who had previously taught at Edgbaston High School and at Notting Hill High School (GPDSC) (Lilley, 1981: 18). After moving to Fitzroy Square in 1885 the college was renamed the Maria Grey College.

The area of science teaching, and particularly chemistry, seemed to be one specific focus of the college. In its description of Maria Grey College's new building, opened in 1892, the *Maria Grey College Magazine* refers to 'the laboratory with every possible aid to the study of chemistry ...' (Anon., 1892: 5–6). There were periodic mentions of the teaching of science at the college. For example by 1897, chemistry teacher Edna Walter (see online Appendix 2) was giving occasional lectures on 'Science Teaching in Girls Schools' (Anon., 1897: 2). In 1902 it was noted that: 'Miss Fortey was teaching science for the Summer Term' (Anon., 1902: 2–3). This was presumably Isabel Fortey (see online Appendix 2) who vociferously supported the need for academic chemistry for girls (see chapter 4).

There was a whole paragraph on science teaching in the Maria Grey College Annual Report the following year: 'A few Science students also had the very great advantage of going regularly once a week during the last two terms to hear a Course of Science Lessons given by Miss Janau in the Central London Foundation School for Girls' (Anon., 1903: 3). Then in September 1919 Marguerite Muriel Grimes (see online Appendix 2) was hired. Grimes was the first of a series of highly qualified instructors in chemistry teaching.

Teachers' training colleges for women at Cambridge University

Frances Buss and Sophie Bryant of North London Collegiate School both considered it vital for prospective women teachers to have teacher training in addition to an academically strong background. In their opinion it made sense to have the training college located in Cambridge, convenient for the graduates of Girton and Newnham. The new institution was approved and the Cambridge Training College (CTC), later named Hughes Hall, was opened in 1885 (Martin, 2011). In September of 1886, Ida Freund was appointed as Lecturer in Method (Bottrall, 1985: 9), though there is no indication that she taught any chemistry. However, in 1900, one of the science teachers at North London Collegiate, Clotilde van Wyss, was hired as part-time science lecturer at CTC (ibid.: 40). The science facilities at the CTC consisted of one room, and it was reported that in 1922 the size of the science room had been halved to make more room for student accommodations (ibid.: 62). Only in 1937 was a properly equipped science laboratory constructed (ibid.: 73).

Nine years later the CTC was joined by a second women's teacher training institution at Cambridge, that of Homerton College. Homerton Academy had been founded in London in 1695, then re-founded as a teachers' training college in 1850. With the move to Cambridge in 1894, it became a women-only institution (Simms, 1979). From 1903 Homerton College had chemist Hilda Jane Hartle on its staff. Hartle, whom we mentioned previously in chapter 4, was born in 1876 and educated at KEVI. She entered Newnham College in 1897, completing the Natural Science Tripos in 1901. Moving back to Birmingham, Hartle spent two years as a research chemist with Professor Percy Frankland at the University of Birmingham. She taught chemistry at Homerton College until 1920, when she accepted a position as principal of the Brighton Municipal Training College for Teachers. Hartle remained there until her retirement in 1941. She died in 1974, age 98 (Rayner-Canham and Rayner-Canham, 2008: 479–81).

The life of women teachers

The large majority of women chemistry teachers stayed single. To them, teaching science to girls was their lifelong career path. An editorial of 1932 in the *Journal of Education* remarked: 'Entering the teaching profession for a woman is equivalent to entering a nunnery; for few marry. Their colleagues look upon those who get engaged as deserters of the cause; not that teachers have many opportunities of meeting the right sort of men' (Anon., 1932b: 560–1). Marriage, with a few exceptions, spelled the end of a teaching career.

In an article in *The Girl's Realm*, Dorothea Beale laid out her principles of what it took for a girl to be a teacher:

... a teacher must keep in the current of thought; if she does not, her mental powers deteriorate, as her muscular system does when she takes no exercise, and then her lectures become dry, stale and uninteresting ...

A teacher must be content to deny herself much that she might do her work without distraction; to those who do this the promise is yet fulfilled – they have a hundredfold more in this present life – peace, content, sympathy, the joy of feeling that they have not lived in vain, and the hope that they may one day dare to say, "I have finished the work Thou gavest me to do."

(Beale, 1900: 623)

Most of the schools that we discuss were day schools and the depressing life of the day school teacher was described by Beatrice Orange in her discourse on 'Teaching as a Profession for Women', in the first volume of *The Woman's Library*. This gloom-filled account included the options for accommodation for the woman teacher:

Here we must touch on one of the problems in the life of the day school teacher – where to live, both cheaply and comfortably. Mistresses' houses are almost invariably failures; flats are expensive and a source of anxiety, for the teacher's tenure of office is apt to be dependent on the whim of a headmistress, and in any case she never reckons on permanently remaining in one place; and so the assistant mistress usually solves the difficulty by choosing the admittedly uncomfortable, but undeniably convenient lodging, which can be left at short notice, and where service, such as it is, is supplied. Nevertheless, to most people, it is a distinct hardship to live for weeks and months together in a small house in a mean street, in an atmosphere of woollen mats and chromo-lithographs ...

(Orange, 1903: 94)

Then Orange described the 'loneliness and dulness [sic]' of such a life:

For the teacher labours under the disadvantage of coming into contact with no one but her pupils and her fellow-teachers during the day; thus differing from the doctor or nurse or factory inspector [other careers for women], all of whom meet with a great number of both men and women in the course of their work; and in her spare time she has to choose between the society of her fellow-teachers or her own. For, save in exceptional circumstances, young, single working women are practically excluded from society ...

(ibid.: 95)

Orange also noted the pressure for career advancement:

> It seems to be the fashion at the present time in most day schools to appoint young headmistresses; and the assistant mistress who has not secured a post as head by thirty-five has need to feel anxious. Should she be forced to seek a new post even as assistant mistress after this age, she would not find the matter very easy; the teaching profession being one where youth counts for more than experience. Therefore, again, I would urge the advisability of emigrating before it is too late to adapt one's self to new conditions, and of seeking in new countries what may still be had for the asking.

(ibid.: 78)

For teaching at both day and boarding schools there could be considerable pressure from the headmistress not to be emotionally and/or physically exhausted – and this in an age of such virulent diseases as tuberculosis, and, later, Spanish influenza. As an example, the biographer of Olive Willis, first headmistress of Downe House remarked: 'Olive had extraordinary physical stamina, and was apt to expect the same in her colleagues ...' (Ridler, 1967: 104). It comes of little surprise to find some of the teachers at this and other schools having retired comparatively young due to 'ill-health'.

Some of the women chemistry teachers

There are many accounts of the pioneering headmistresses, but the teachers – in particular the chemistry teachers – have vanished from the record. Yet it is these young women who brought their enthusiasm for chemistry to the subsequent generations. Some specific biographies are sprinkled through this book in relevant locations; however, many more are provided in online Appendix 2. Here we give an overview.

Many of the women chemistry teachers stayed at one school for most of their lives. Others seemed to have 'flitted' from school to school. The cause – if there was any general cause – of transient positions is unclear. However, it would seem that, at least during the First World War, there was a significant shortage of women chemistry and physics teachers.

The headmistress W.M. Kidd, at the Girls' Grammar School, Maidstone, gave a presentation on the supply of women teachers in secondary schools for girls at the 1916 Conference of the Association of Headmistresses that was reprinted in *School World*. She reported that, whereas there was a glut of women teachers in such subjects as English and history, there was a shortage in science:

> The temporary dearth of teachers of mathematics and science (especially chemistry and physics) is really serious. This, of course, may right itself after the war. But at the present time a great many [physical science] mistresses are entering boys' schools, others are becoming analysts and instrument-makers and taking posts in chemical and physical research laboratories. In such posts the hours are shorter and the evenings not burdened by corrections. Then, again, the war is creating an ever-increasing demand for women doctors.
>
> (Kidd, 1916: 267)

The 'missing' earliest women chemistry teachers

The record for women chemistry teachers is sadly lacking for the earliest years. In particular, though the GPDSC schools have religiously kept their staff registers, the pages did not commence until around 1900. As an example of this missing period, Jane Sayers, the school historian for Notting Hill and Ealing High School commented:

> Mrs Grey [headmistress at Notting Hill High School] had maintained that Science should be part of the curriculum. In 1887 Miss Edith Aitken [see chapter 3], a Girtonian arrived to teach Science: she stayed until 1892. Miss Adamson [see online Appendix 2] returned in 1890 and was to produce some very remarkable pupils, but the question arises as to who taught Science in the first decade, who in fact had taught Mary Adamson between 1878 and 1882 and Beatrice G. Taylor between 1876 and 1880? The answer may be Mrs Pilcher who took her BSc in 1883.
>
> (Sayers, 1973: 103)

All that we could discover of 'Mrs Pilcher' is that as Sophia Margaretta Lewis she had begun teaching at Notting Hill High School in 1877. In 1883 she obtained a BSc from University College, London (ibid.: 84), then in 1884 she resigned in order to marry a Mr G.T. Pilcher.

A similarly enigmatic figure was Miss Clarke of Berkhamsted School for Girls. In the school history Garnons Williams noted:

In her early days she [Miss Disney] had employed as a science teacher a member of the Boys' School staff, the Rev. A. Cole, but in 1896 she substituted a full-time Cambridge woman graduate, a Miss Clarke from Newnham. This seemed to be making history; Miss Clarke was not only the first scientist but also the first graduate on the staff. She was of unimpeachable credentials, having been at school at Cheltenham under Miss Beale ... She seems to have been a successful and even inspiring teacher, but unfortunately her career was interrupted by illness and on her recovery she resumed it elsewhere.

(Williams, 1988: 28)

Not even in the *Newnham College Register* could any additional information on Miss Clarke be found.

Some Women Chemistry Teachers

In chapter 2 we provided backgrounds on some of the chemistry teachers at the Quaker schools; then in chapter 3, on the chemistry teachers at NLCS and CLC. We have been able to piece together brief biographical accounts for several other teachers. The detailed information is provided in online Appendix 2. However, we feel it important for the benefit of the reader to list all the women chemistry teachers we have included and the schools at which they taught for lengthy periods. Unless stated otherwise, the biographical information is in online Appendix 2.

Adamson, Mary M.: Princess Helena College; Bromley High School; Notting Hill High School; Portsmouth High School.

Aitken, Edith: Manchester High School for Girls; Nottingham Girls' High School; Notting Hill High School; North London Collegiate School (see chapter 3).

Auld, H.P.: St Leonards School; Esdaile School (see chapter 10).

Beard, Miss: Park School, Glasgow (see chapter 10).

Benn, Ethel; St Leonards School.

Birt, Margaret Ethel: Brighton and Hove High School; St Paul's Girls' School.

Broadhurst, Miss: Park School, Glasgow (see chapter 10).

Burdett, Miss: Park School, Glasgow (see chapter 10).

Cam, Marjory Thekla: Oxford High School; Bromley High School (see chapter 8).

Carter, Maude Sampson: County School for Girls, Pontypool (see chapter 9).

Coignou, Caroline Pauline Marie: Pendleton High School for Girls; Manchester High School for Girls.

Collier, Kathleen Mary: Bedford High School.

Dalston, Daisy Florence: Sheffield High School; Streatham Hill High School.

Davies, Dilys: North London Collegiate School (see chapters 3 and 9).

Davis, Ivy Rosina: Howell's School, Llandaff; Howard Gardens Secondary School.

Dove, Jane Frances: St Leonards School; Wycombe Abbey School (see chapter 10).

Dunbar, Mary: Edinburgh Ladies' College; Chesterfield High School for Girls.

Fortey, Isabel Comber: Park School, Glasgow; King Edward VI High School for Girls, Birmingham; Newnham College, Cambridge; Sheffield High School; Cambridge Training College for Women; Private school, Greenock; St Hilda's School, Ootacamund, India; Diocesan College, Calcutta.

Gibson, Florence: Cardiff High School for Girls (see chapter 9).

Grimes, Marguerite Muriel: Boys' Grammar School, Hitchin; Brondesbury and Kilburn High School for Girls; Maria Grey Teachers' Training College.

Harker, Katherine Nancie Helen: Leigh Girls' Grammar School; North London Collegiate School.

Heath, Annie Grace: North London Collegiate School (see chapter 3).

Heather, Lilian Frances: Downe House School.

Hunt, Annette Dora: Weymouth High School; Croydon High School; Sutton High School.

John, Margaret E.: King's High School for Girls, Warwick; Cardiff High School for Girls.

Leeds, Kathleen Mary: Portsmouth High School; Croydon High School.

Lees, Edith Sarah: Clapham High School.

Lewis, Iva Gwendoline: Roedean School; Wimbledon High School (see chapter 8).

Leyshon, Eluned: Glanmôr Girls' Secondary School, Swansea (see chapter 9).

MacDonald, Evelyn: Wheelwright Grammar School, Dewsbury; Oxford High School.

O'Flyne, Norah Dorothy: County School for Girls, Pontypool (see chapter 9).

Patterson, Dorothy Christina: Wheelwright Grammar School, Dewsbury; Perse School.

Quartly, Lilian Ada: Clapham High School.

Raymond, Yolande Gabrielle: Liverpool (Belvedere) High School; Sydenham High School; St Paul's Girls' School; Kidderminster High School (see chapter 5).

Ready, Gertrude Elizabeth: Rochester Grammar School for Girls; Nottingham Girls' High School.

Rich, Mary Florence: Howell's School, Llandaff; Grantham Ladies' College; Roedean School; Granville School, Leicester (see chapter 9).

Rippon, Dorothy May Lyddon: Downe House School.

Stern, Rose: Bangor County School for Girls; North London Collegiate School (see chapter 3).

Stopard, Winifred Mary: Glossop Grammar School; Central Secondary Girls' School, Sheffield; Sheffield High School (see chapter 8).

Sutton, Mary Winifred: Redland High School; St. Martin-in-the-Fields High School (see chapter 7).

Taylor, Clara Millicent: Clapham High School; St Paul's Girls' School; Northampton School for Girls; Redland High School for Girls.

Taylor, Millicent: Cheltenham Ladies' College (see chapter 3).

Thomson, Jean G.: St George's School; Falkirk High School (see chapter 10).

Turner, Margaret K.: Pontypool County School for Girls (see chapter 9).

Virgo, Muriel Elizabeth: Wakefield High School; Streatham Hill High School; Cheltenham Ladies' College; Ipswich High School.

Walter, Lavinia Edna: Central Foundation Girls' School.

Watson, Mary: Cheltenham Ladies' College (see chapter 3).

Willey, Emily Mary: Chelmsford County High School; Twickenham Girls' High School; Girls' High School, Stafford; Howell's School, Llandaff.

Winny, Alice Georgina: Howell's School, Llandaff (see chapter 9).

Younie, Elinor M.: Park School, Glasgow; Central Secondary School for Girls, Sheffield; St George's School.

Figure 5.2: Caroline Coignou, chemistry staff member at Manchester High School for Girls, assisting students with their chemistry laboratory reports, 1905

Source: Archives, Manchester HSG

Association of Women Science Teachers

Often forgotten in educational history, the Association of Women Science Teachers (AWST) played an important role in linking together these pioneering women (AWST Minutes).

Origins of the AWST

It was the science section of the London branch of the Association of Assistant Mistresses (AAM) from which the AWST was founded. In the autumn of 1911 there was a consensus that it was important to have an organization to which all women science teachers could belong, not only those who were eligible as members of the AAM. There was deemed to be a particular urgency as, in their view, science teaching was undergoing rapid changes.

Thus a formal body was needed that could discuss issues and present the conclusions to the educational authorities as the collective viewpoint of its members. Membership would be open to all current and former teachers of science at the secondary or post-secondary level. A six-member committee was struck to draft proposals for such an organization (Layton, 1984: 39–40). Of the six members, three were chemistry teachers: Edith Lees of Clapham High School (see online Appendix 2); Rose Stern of North London Collegiate School (see chapter 3); and Yolande Raymond of St Paul's Girls' School.

Yolande Gabrielle Raymond, the third chemistry teacher on the committee, was born in 1871 in County Kerry (White, 1979: 107). She was educated at the Clergy Daughter's School in Bristol, entering Newnham College in 1890. Raymond completed the Natural Science Tripos in 1893, subsequently holding positions as chemistry teacher at Liverpool (Belvedere) High School (1893–8) (Anon., 1898: 4); Sydenham High School (1898–1905); and St Paul's Girls' School (1905–12) (Anon., 1912: 2). She was appointed headmistress of Kidderminster High School in 1912, a post she held until 1932. Raymond died in 1952.

Information about the proposal for the organization was circulated in January 1912 and by the end of March 1912, 103 women science teachers had joined (AWST Minutes). The first general meeting was held in November 1912 at the London Day Training College, and the minutes report that the meeting was followed by an address given by Henry Armstrong entitled: 'Science befitting girls'. Unfortunately, no record of the lecture survives either in the AWST archives or in Armstrong's published public lectures.

The organization was initially open to both female and male members, and its original name was the Association of Science Teachers. In reality, as discussed below, it was an essentially women-run organization from the beginning, thus we use its later name of the Association of Women Science Teachers throughout. Membership of the AWST was also open to women teachers from post-secondary institutions. One of those to join was Hilda Hartle (see above).

Two meetings were held each year: a winter one in London and a summer one elsewhere in the country. During the early years, meetings were often followed by talks or tours. For example, in 1914, the biochemist Ida Smedley Maclean (see online Appendix 3) spoke on 'The Formation of Fats in Living Organisms'.

In 1915, an extraordinary meeting was held to discuss the syllabi in each science for girls of 16 and 18 years of age (Anon., 1915). The authors for the chemistry syllabi were Rose Stern and Dorothy Marshall (see above), who at that date was chemistry teacher at Clapham High School. Unfortunately, the syllabi were not found in the AWST records.

The AWST in the 1920s and 1930s

The first chemistry teacher to occupy the position of president had been Edith Lees in 1919 (Anon., 1962) (see online Appendix 2). Then for 1912 to 1913, the vice-president was Rose Stern. Women chemistry teachers continued to be prominent in the AWST in later decades.

The president in 1921 and 1922 was also a chemist, M. Beatrice Thomas (see above) of Girton College (Taylor, 1954: 43–4). Clara Taylor of Redland High School (see online Appendix 2) – not to be confused with Millicent Taylor of Cheltenham Ladies' College of chapter 3 – was president for 1925 and 1926 and vice-president for 1927; Margaret Birt of St Paul's Girls' School (see online Appendix 2) was honorary secretary from 1926 to 1935; Margaret Tomkinson (Butler and McMorran, 1948b: 652–3) was president for 1932 and 1933; while Mary Winifred Sutton of St Martin-in-the-Fields High School was membership and meetings secretary in 1935, and president 1949–50.

The AWST was initially a very inclusive organization, and though men were welcomed, few ever joined, preferring the exclusive men-only Association of Public School Science Masters, renamed the Science Masters' Association (SMA) in 1919 (Jenkins, 2013: 1–29). The SMA published a journal, the *School Science Review*, and the executive of the AWST saw the SMA journal as a valuable printed means of interacting within the AWST membership. In the negotiations that followed, the SMA insisted on retaining control of the journal but permitted the AWST to submit contributions. In return, the SMA insisted that the AWST inserted the word 'women' in its name, to prevent the SMA losing membership to the previously open membership of the AWST. Thus in 1922, it was proposed by Hilda Hartle, and approved unanimously, that the name officially be changed from the Association of Science Teachers to the Association of Women Science Teachers. At that time, the AWST had a membership of 360 members compared with 729 for the men's SMA.

Of the 465 members of the AWST in 1926, only ten were married or widowed. David Layton has contended that the AWST played an important social as well as academic role in the women science teachers' lives:

> For the large proportion of unmarried women members, the Association's social functions were as important as, and probably indivisible from, its professional ones. The 'cozy intimacy' of the meetings in the early years, and the 'jolliness and earnestness' of later gatherings were much valued by the predominantly spinster membership.
>
> (Layton, 1984: 63)

The AWST Report of 1921

The AWST Annual Report of 1921 was of particular interest as the focus was chemistry and the discussions at the annual meeting were given in full. To begin, Clara Taylor gave a lengthy annual address in which she first defined the girls taking science:

> ... that while I am not excluding from what I have to say, the few girls who have special aptitude for Science, I am thinking chiefly of the mass of reasonably intelligent girls whose brains are not of the creative order, those who have no great power of mental initiative, but whose intelligence can learn and appreciate a great many facts and methods that come within the scope of natural science ... It is this group who form the majority of the girls to whom we teach Science in School.
>
> (Anon., 1921: 4)

Then Taylor critiqued Armstrong's heuristic method (see chapter 4):

> We have not made the most of our resources, for so often we have deliberately stood aside and tried to let the facts speak for themselves; and to most of us facts will not speak for themselves; they need an interpreter, and we have not sufficiently realized that our position is to interpret.
>
> (ibid.: 6)

Taylor's address was followed by a presentation by Rose Stern on the teaching of chemistry in the middle school. Stern began by giving a general overview of the purpose of the chemistry laboratory as an exercise in developing mental powers:

> A great deal has been accomplished when a whole class of beginners can set up an experiment, bending their own glass tubing, and fitting corks so that it succeeds. How often does one see a girl trying to fit a large cork into a small opening, and how often is there disappointment on finding out that a cork becomes larger when a tube is put through it. These may be small things, but they ensure good training. The class learns that little faults lead to bad endings.

(ibid.: 8)

Stern then presented her traditional approach in teaching chemistry, as published in her book: *A Method of Teaching Chemistry in Schools* (Stern and Hughes, 1906), followed by her later approach using the history of chemistry. In order to appreciate what chemistry was taught to girls in that period, the two discourses are given in online Appendix 7. It is of note how much of the courses are experiment-based and taught following Armstrong's heuristic approach of students discovering chemistry for themselves. Also, the chemically knowledgeable reader would be aware that many of the experiments described would never be permitted today on safety grounds.

Branches of the AWST

The science section of the London branch of the AAM (which had founded the AWST) continued in existence as the London branch of the AWST (Layton, 1984: 50–4). As the AWST membership was largely centred on the London region, there was a considerable overlap with the London branch; for example, Rose Stern presided over the London branch for the first three years. Among the talks presented at the London branch meetings were: 'Science for backward girls'; 'The training of science teachers'; and 'Reasons for the present shortage of women science teachers' (ibid.: 51). In addition, periodic excursions were organized, including one in 1916 to Baird and Tatlock, laboratory chemical suppliers and scientific instrument manufacturers at Walthamstow.

Autonomous branches were also set up across the country, the earliest being: North-West (1914); Midlands (1918); Wales (1921); North-East (1926); and Northern (1926). An account of the convoluted history of the North-East (Yorkshire) branch has been given by Edgar Jenkins (Jenkins,

2009) while the Welsh branch will be discussed in chapter 9. There was a considerable degree of autonomy among the branches of the AWST, in part because local meetings were easier to organize and attend.

In 1932 the North-West branch of AWST developed a secondary school curriculum for science (Anon., 1932c), to be used between the ages of 11 and 15. Though it is not clear what became of the proposal, the chemistry component gives a sense of the teaching of chemistry in that era. The section on the Chemistry Scheme for Post-Primary Schools is listed in online Appendix 8.

These first generation teachers were the women who inspired the young girls in their chemistry classrooms, exciting them about molecules and chemical change, as we show in the next chapter. In doing so, most of the women teachers devoted their lives to the cause of science teaching, and chemistry teaching in particular.

Chapter 6

Practical chemistry at girls' schools

Trials of the "Lab"
The smells of the "Lab" are the greatest of trials
Ignorant voices beg
That you cease from making ammonia gas
And the smell of a rotten egg

(Anon., 1922: 1012)

To teach chemistry required a chemistry laboratory. From the mid-1870s the headmistresses of many independent girls' schools worked diligently to provide chemistry facilities for their students. Generally the early girls' schools occupied large private houses and initial laboratory spaces were modified attic or basement rooms. Most of the schools then constructed adequate laboratories in the period 1890–1910 and replaced them with custom-built laboratories in the 1920s or 1930s.

Here we look at these three time frames in the context of a demand for a chemistry laboratory. It is the interest and enthusiasm by the girls for practical chemistry that becomes so apparent in all the accounts. For example, in 1906, Gerty Turner in Upper IIIb at the Central Foundation Girls' School reported in the school magazine: 'We are now learning experiments in Physics and Chemistry, which we all like very much, and we feel rather big in our holland overalls, that we wear for Chemistry' (Turner, 1906: 251).

However, first we need to discuss why the provision of a chemistry laboratory was so important for a girls' school.

The need for a chemistry laboratory

There seem to have been three sources of pressure for a chemistry laboratory: that chemistry and the accompanying practical laboratory was essential in order to be considered 'as good as the boys' schools'; that practical chemistry taught valuable life-lessons; and that practical chemistry was required for admission to university science degrees.

To be as good as the boys' schools

Some of the headmistresses regarded chemistry, and a chemistry laboratory in particular, as a defining feature of an academic girls' school. For example, Frances Dove, when she founded Wycombe Abbey School, considered a chemistry laboratory of utmost importance, as author of the school's history, Lorna Flint, recounted:

> But the dearest of all projects to Miss Dove was the provision of a laboratory. Her first endearingly wistful reference to it occurs in her Report to the Council in 1902, where she confesses that she is 'longing for the time when Practical Physics and Chemistry can find a place' among the School's subjects ... the Minutes of May 1907 record that a laboratory was 'urgently needed'.
>
> (Flint, 1989: 19)

And in October that year, the construction of a chemistry laboratory was finally authorized.

Figure 6.1: Young students at Dame Alice Harpur School, Bedford, performing chemical experiments, undated
Source: Archives, Bedford HS

There seems to have been widespread competitive pressure to install chemistry facilities at girls' schools around the 1900s. This phenomenon was noted in the history of Dame Alice Harpur School written by Constance Broadway and Esther Buss:

> A strong movement towards the improvement of science teaching in girls' schools developed during these years [about 1900], and

in 1904 the Science and Art rooms were fitted up. The provision of proper benches, sinks, the familiar Bunsen burners and glass apparatus made possible the first experimental work ...

(Broadway and Buss, 1982: 34–5)

Life-lessons from practical chemistry

Between 1899 and 1904, the girls' magazine *The Girl's Realm* published a series of articles on famous girls' schools. In many of these articles, the author would comment upon the chemistry facilities. The most prolific author of such school reviews was Christina Gowans Whyte. As an example, for Mary Datchelor School she wrote: 'Upstairs two well-equipped laboratories for chemistry and physics are included amongst the class-rooms' (Whyte, 1901–2: 270).

Whyte waxed lyrical about practical chemistry and its role in a girl's education in her article in 1900 on Roedean School:

> Science has a laboratory, fully equipped in every respect for the course overtaken in the school curriculum. The mysteries, amounting almost to terrors, which the imagination of a girl sometimes conjures up at the idea of chemistry, may fade from her mind when she sees the white room and the dainty, spotless appearance of tubes and bulbs and blowers in that sanctum. People have a way of believing that such studies, as for instance the study of chemistry, are only useful to the bookworms of women, who intend to take university degrees. The girl at Roedean School, no doubt, already discovers the great practical advantages to be derived from the study of chemistry. For one thing, there are photographers among them, and the world would have been spared some disastrous results in amateur photographs by ladies if chemistry had been at all times a necessary part of their education.
>
> (Whyte, 1900: 1063–4)

It was also believed that teaching rigorous science, especially chemistry, with its inherent dangers, was a fitting training for life. This view was eloquently expounded by Whyte in her article on Bedford High School in *The Girl's Realm*:

> Practical work in chemistry is included in the curriculum. Practical chemistry is one of the best among modern educational improvements. It teaches things which go much deeper into our consciousness than mere words could ever go. It teaches

consequences – the stern, certain consequences of doing quite the right or the wrong thing. It never makes a mistake, or slurs over a little bit of carelessness, or pretends everything is right when everything is the reverse. And for girls who have to go through with life, it is not a bad thing to learn when young to expect the natural consequences of an action, even to the correct or incorrect testing of a compound or simple liquid.

(Whyte, 1900–1: 872)

Figure 6.2: The attic chemistry laboratory at Bedford High School for Girls, 1902

Source: Archives, Bedford HSG

Entry requirements for universities

Several accounts mention the need to meet university regulations for admission to science degree programmes. In an 1884 article on science teaching in girls' schools published in the *Journal of Education*, the anonymous author stated:

With a small amount of material, and with no more space than is afforded by an ordinary school-room, much may be done to make Chemistry, as it should be, the basis of all Natural Science teaching. Where a chemical laboratory, however small, can be obtained, it becomes possible, as well as desirable, that Chemistry should be taught more thoroughly and practically. Take, for instance, such a course of Chemistry as that prescribed for the London Matriculation Examination. The facts and phenomena should be taught first by the aid of experiments performed by the teacher. This should then be followed up by practical lessons, in which the pupils themselves perform the experiments. The simple gases – Oxygen, Nitrogen, Hydrogen, &c. – can be prepared by a class of pupils without difficulty in a moderately sized laboratory, and students who have themselves actually performed such experiments acquire a knowledge of the laws of chemical reaction, and of the properties and constitution of matter which would be impossible without such means.

(Anon., 1884a: 353)

At Notting Hill High School, a lecture room with a small laboratory was constructed in 1877 for students hoping to enter London University (Jones, 1885: 5). The laboratory must have been more of a chemical preparation facility, for one of the students later reminisced at the 1911 winter meeting of the Notting Hill High School Old Girls' Association:

The Lecture-Room ... was always most useful for special lessons, there was no Laboratory in those days, ... and there the girls were prepared in Chemistry for the London Matriculation, and they generally passed too! Of course they could do no practical work, the experiments were performed before them.

(Anon., 1911: 29)

The presenter added: 'The last addition to the buildings was the Laboratory in 1898. It was much needed and it was a great relief to Form V Upper to get rid of all the bad smells made by the Chemistry Class' (ibid.: 31). This was actually a combined lecture-laboratory (see Figure 6.3).

Figure 6.3: Notting Hill High School chemistry room, undated

Source: Archives, Notting Hill HS

At St Swithun's School, too, it was the university entrance exams that provided the stimulus for practical chemistry, Edith Finlay recalled:

> In 1897, four of us entered for our first practical exams. in the Joint Board Certificate. In those days, a 'don' was in charge in cap and gown. An enterprising examiner had given red phosphorus as the unknown substance. About ten minutes after we had commenced, a nervous candidate dropped a glowing match on the 'unknown' – result, a wild flare and we all 'knew'. Hardly had the examiner extinguished this when it was discovered that a pile of dusters was on fire; this in turn was extinguished. Then suddenly the bottom came out of a medicine bottle improvised to contain sodium hydrate [sodium hydroxide], devastating a varnished table and all the candidate's papers. Wearily our friend came for the third time to the rescue, remarking, "My life is insured – I only hope yours are!" Possibly we had more thrills out of our Science lessons in those days than under modern conditions. To me at least they are a real joy to recall …
>
> (Finlay, 1934: 20)

A chemistry laboratory as a status symbol

A chemistry laboratory was not only regarded as a necessity for one or more of the reasons listed above, but it was also a symbol of the modernity of the school and a source of pride.

Keeping up with the 'top' girls' schools

In 1898 a compilation titled *Work and Play in Girls' Schools by Three Head Mistresses* was published (Beale *et al.*, 1898). The three headmistresses were Dorothea Beale of Cheltenham Ladies' College (CLC), Lucy Soulsby of Oxford High School (a GPDSC school), and Frances Dove of Wycombe Abbey. This work, containing contributions by experts in each subject area, was intended to define the standards expected at independent girls' schools – including chemistry. The chapter on 'The Teaching of Chemistry' was authored by Clare de Brereton Evans.

Clare de Brereton Evans had been educated at CLC and had obtained an external BSc (London) degree in 1889 while still at the CLC. She then undertook research with Professor Henry Armstrong at the Central Technical College, which resulted in her being the first woman awarded a DSc degree. From 1904 until 1912 De Brereton Evans was a lecturer in chemistry at the London School of Medicine for Women, while also being a researcher with

Sir William Ramsay at University College, London (Rayner-Canham and Rayner-Canham, 2008: 159–60). De Brereton Evans was convinced that junior, as well as senior girls needed exposure to practical chemistry:

> For success in examinations it is now necessary to have a certain amount of practical knowledge of chemistry, and examination classes are therefore given some practical training, but this reform still remains to be extended universally to the junior classes, which need even more than the senior ones that the teaching should be objective: a child may learn and repeat correctly a dozen times that water is composed of oxygen and hydrogen, and the thirteenth time she will assure you that its constituents are oxygen and nitrogen; but let her make the gases for herself, test them and get to know them as individuals, and mistakes of this kind will become impossible.
>
> (De Brereton Evans, 1898: 310–11)

Claimants of priority

The network of Girls' Public Day School Company (GPDSC) schools were aware of each other's laboratory facilities. However, among the other independent girls' schools many believed that their own school was one of the very rare ones to pioneer the teaching of chemistry and to have a chemistry laboratory.

For example, in the history of St Paul's Girls' School, Hammersmith, we note that: 'In 1904, St Paul's was one of the few girls' schools in Great Britain to pay serious attention to Science education. Most high schools had no laboratories and almost ignored the Sciences, though there were signal exceptions' (Bailes, 2000: 47). Similarly, in the history of Milton Mount College, we read: 'In many aspects Milton Mount was a pioneer among girls' schools. [In 1883] It was the first to have a laboratory …' (Harwood, 1959: 13).

We show here that, in fact, most of the independent girls' schools we visited had installed chemistry laboratories in the 1880s, 1890s, or at the latest in the early 1900s.

Comparing chemistry laboratory facilities

School parties visited other schools and reported on their chemistry laboratories. For example, in 1887, a group of senior girls from Milton Mount College visited North London Collegiate School (NLCS). In the account of the visit in the *Milton Mount Magazine*, it is commented: 'The laboratory! In the laboratory we remembered just in time the words, "Thou shall not covet thy neighbour's goods" – nor his laboratory, we mentally supplied for ourselves' (Anon., 1887a: 142).

Similar scrutiny was undertaken when groups from girls' schools visited universities and colleges. For example one of Rose Stern's students at NLCS reported how well equipped the NLCS lab was compared with that of Royal Holloway College (RHC) during a school visit there in 1897. Greta Hahn wrote in *Our Magazine: North London Collegiate School for Girls*: 'A hasty peep into the laboratory showed that it was very similar to our own school laboratory, but, as someone triumphantly remarked, much more untidy ...' (Hahn, 1897: 109).

Similarly a group of students from Manchester High School for Girls (MHSG), visiting Cambridge in 1901, were of the opinion that their chemistry laboratory was superior to that of Newnham College. Writing in the *Magazine of the Manchester High School for Girls* we read: 'The Newnham Chemistry Laboratory was inspected on Monday morning, but was agreed to compare very unfavourably with the Chemical Laboratory of the Manchester High School, whatever the standard of work may be' (McNicol, 1901: 48–9).

Practical chemistry in novels for girls

The importance of the possession of a chemistry laboratory even extended to fiction. We described in chapter 1 how a new genre of fiction for girls defined the identity of the New Girl. The descriptions of the semi-fictional girls' schools were designed to show that the schools themselves were progressive and up-to-date. Such an image required that the school offered chemistry and, in particular, contained a chemistry laboratory.

The author who exemplified the New Girl stories was Angela Brazil (Freeman, 1976). Several of Brazil's stories included mention of a chemistry laboratory. For example, in *The New Girl at St Chad's*, the New Girl, Pauline, noted that:

> The buildings were modern and up-to-date, with all the latest appliances and improvements. They were provided with steam heat and electric light; and the gymnasium, chemical laboratory, and practical demonstration kitchen were on the newest of educational lines.
>
> (Brazil, 1911: 39)

Similarly, in *A Patriotic School Girl*, the narrator, Marjorie, observed on arriving at her new school that: 'It had been built for a school, and was large and modern and entirely up-to-date. It had a gymnasium, a library, a chemical laboratory ...' (Brazil, 1918: 23). Even an old school setting needed to be modernized by a chemistry laboratory. For example, in *The School by the Sea*, the school, Dower House, is described as a former nunnery founded in the fourteenth century. The headmistress, Miss Birks, muses that:

> Could the Grey Nuns have but returned and taken a peep into the well-equipped little chemical laboratory, they would probably have fancied themselves in the chamber of a wizard in league with the fiends of darkness, and have crossed themselves in pious fear at the sight of the bottles and retorts ...
>
> (Brazil, 1914: 26)

Was there any special reason why Brazil referenced chemistry and chemistry laboratories? Brazil attended Ellerslie School, near Manchester. Brazil's biographer noted: 'Miss Millington, her headmistress [at Ellerslie], was an enlightened educationalist with many of the qualities that immortalized Miss Buss, the famous founder of the North London Collegiate and Camden Schools' (Freeman, 1976: 23). We find no hint in Brazil's autobiography of the origin of Brazil's periodic references to chemistry in her novels; the sole brief mention of science relating to her junior days was 'an amiable clergyman who came weekly to give little lectures on science ...' (Brazil, 1925: 160). Perhaps her image of an up-to-date school had come from her sister, Amy. Amy attended MHSG, which had possessed a chemistry laboratory since 1881.

Earliest practical chemistry at girls' schools

The first school to offer practical chemistry in a laboratory setting was claimed to be the (boys') Manchester Grammar School in 1868 (Perkin, 1900). The girls' schools were not far behind. The earliest we could find for independent girls' schools was the NLCS in 1875 and CLC in about 1876.

Practical chemistry in the classroom

We mentioned above that at Notting Hill High School, chemistry demonstrations were initially performed in a lecture room. At some of the schools, students were permitted to perform experimental chemistry work in classrooms. This was the case at Ipswich High School. Louisa Frost, a student from 1898 to 1901 recalled:

> We were encouraged to "do it yourself" and, as General Elementary Science was compulsory in those days and we had no lab., we ... burned our hair (long) and nails, ... made oxygen and hydrogen, and discovered that Epsom Salts was only $MgSO_4\ 7H_2O$!
>
> (Anon., 1978: 8)

Similarly, St Swithun's School initially lacked proper laboratory facilities, but the students took matters into their own hands to encourage the authorities to add a chemistry laboratory:

> In 1893, Miss Fletcher was appointed Science Mistress ... But still we had no [Chemistry] laboratory. Well do I remember these class-room chemistry lessons, where smelly experiments such as the preparation of chlorine had to be hastily pushed outside on the window-sill, and my joy when I was promoted to the honour of cleaning up the mess ... Apropos of our difficulties, I may mention that when in 1895 the question of a new addition to the building and possibly a laboratory was mooted, the Chemistry class prepared an energetic flask of chlorine mixture and left it where the Visiting Committee could not fail to be conscious of its objectionable odour – whether it influenced their decision I do not know, but to our joy we heard the laboratory was to be built.
>
> (Finlay, 1934: 20)

And indeed the science laboratory was opened in 1896, though it was not ideal: 'Our work was still carried on under difficulties, as classes constantly passed through the Laboratory to the small class rooms beyond, and I remember the joy with which we exploded cylinders of hydrogen if possible at such moments' (ibid.: 20).

Figure 6.4: Pen-and-ink sketch of a St Swithun's School student at the chemistry bench, undated

Source: Bain, 1984: xvii

Another school that initially had to 'make do' was The Perse School, Cambridge. One of the students from the 1880s recalled:

My first science lessons were in what would now be considered a makeshift lab.; on the ceiling was a ghastly black mark from a recent explosion in a chemistry class, the tale of which was probably exaggerated, but the evidence on the ceiling struck my heart with terror.

(Anon., 1956: 18)

Laboratory work in chemistry was regarded to be so important at Newcastle High School (NHS) that the girls travelled all the way to Gateshead High School (GHS) for their practical experience:

[about 1895] There was no science laboratory in the school [NHS], and it was arranged that we should have the use of the GHS laboratory. A horse-drawn bus took us there every week directly after morning recreation when we had milk and buns.

(Carter, 1956: 32)

One infers that Armstrong's heuristic method was in use at the GHS, as the account continues: 'Although the Gateshead laboratory was small, our chemistry lessons were on modern lines. We did our own experiments and were taught to find out scientific truths for ourselves' (ibid.: 32).

Basement and attic chemistry laboratories

When the schools were founded, practical chemistry was undertaken in whatever space was available. At South Hampstead High School the first laboratory was in the basement, and it was in use for 20 years until a proper laboratory was built:

[In 1882] The present Chemistry Laboratory did not exist; the only laboratory was the small basement room, now a cloak room, beyond the washbasins. For Science lessons it must have been a dark and dismal abode, but those girls who have a bent towards archaeology may still find traces of the draught cupboard and the one sink and tap ...

The next addition was the new laboratory (our present Chemistry room) which was built in 1902, and which must have seemed palatial and royally lighted after the old dark cell.

(Anon., 1926: 6)

Another school to replace its original laboratory location in the 1900s was Portsmouth High School as was recounted in the *Portsmouth High School Magazine*: 'The school started off [in the 1880s] with one rather small laboratory ... During the early part of 1908, a beautiful science room was added ...' (Anon., 1932: 5).

In some schools an attic room was used as the first primitive chemistry laboratory. This was the case at Belvedere School, as Alison MacKenzie, a student from 1904 until 1912, recalled in the *Belvedere School Liverpool Chronicle*: 'The only Laboratory of any sort [when she started school] was a small top room, with a gas jet and a tap in one corner of the room. However, I was too small then to be allowed into a Lab. and so I may be wrong in these details' (MacKenzie, 1930: 46).

The first chemistry laboratory at Roedean School, too, was an attic room, as Joan Waller recalled in the *Roedean School Magazine*:

> Our first laboratory is now no more than a shadowy recollection of a small room at the top of one of the houses at Sussex Square, where, in our early days, to our great joy and pride, we were allowed to break test-tubes and generally to misconduct ourselves.
>
> (Waller, 1899: 84)

Whereas attic rooms (or basements) as chemistry laboratories were a temporary expedient at most schools, at others they lasted into the 1920s and 1930s. In a history of Alice Ottley School, Valentine Noake described how the attic laboratory was in use into the 1930s:

> A slightly older class would be instructed in setting up the apparatus for preparing oxygen or isolating ammonia gas, and would then go to the benches and fit flasks with rubber stoppers, bend glass tubing to appropriate shapes, pour acids cautiously into thistle funnels, and collect and test the results. It was sheer joy to be allowed to prepare the evil-smelling sulphuretted hydrogen [hydrogen sulphide] in the fume cupboard: it gave the budding chemist a sense of vast and noisome power.
>
> (Noake, 1952: 133)

For the schools of the GPDSC, the Central Council held sway in many matters, including the installation of chemistry laboratories. Though some of the GPDSC schools already incorporated a chemistry laboratory, in 1882 an edict went to all GPDSC schools that any new school building must incorporate one:

> A chemical laboratory should be provided not less than 15 ft wide, and 16 ft to 20 ft long. It should be fitted with the necessary working tables and sinks ... ventilation should be similar to those

in the class rooms, except that two shafts should be provided instead of one. One flue should be provided for ventilating the closet required for producing noxious gases.

(Anon., 1882: 125)

Figure 6.5: A typical GPDSC chemistry laboratory: Brighton and Hove High School, undated

Source: Archives, Brighton and Hove HS

Fees for practical chemistry

In 1884 the council of the GPDSC recommended that a fee was to be charged to all students taking practical chemistry:

> That in all Schools where there is practical teaching in Chemistry, a charge of not less than 5s. per Term shall be made for materials to each pupil doing Laboratory work, and that all breakages shall be replaced by the pupil responsible.

(Anon., 1884b: 3)

In our research we found that the 5s fee (which represented a substantial amount of money at the time) was charged at many – if not all – independent girls schools in that early era. For example the *School Report* of MHSG for 1887 commented that the students were required to pay a 'small charge' for 'supply of materials and apparatus' used in the chemistry laboratory (Anon., 1887b: n. pag.). The MHSG fee would seem to have also been 5s, for at the meeting of the governors in 1888, it was reported that the introduction of the fee of 5s per term had resulted in a reduction in the numbers of girls taking chemistry (Anon., 1888: 251). There was no mention when the fee was terminated.

Figure 6.6: The chemistry laboratory at Manchester High School for Girls, 1905

Source: Archives, Manchester HSG

The fee ceased to be charged at the GPDSC schools in 1907 as a result of the intervention of the Board of Education which sent the GPDSC Council the following instruction:

> The Board do not ... think that the extra fee for practical work in Chemistry done by the advanced Forms in the Course should be continued ... Where Chemistry is taken in the Course, laboratory instruction is a necessary part of the teaching ...
>
> (Anon., 1907: 2–3)

Chemistry lecture theatres

Though in this chapter we are concentrating on the actual laboratory facilities, it is important to note that some schools, such as CLC and MHSG, had both a chemistry laboratory and a chemistry lecture room. The room, more correctly described as a lecture theatre, had tiered student seating and a large well-equipped bench for the chemistry mistress, Caroline Coignou (see online Appendix 2), to perform demonstrations. A note in the *Magazine of the Manchester High School* of 1887 reported: 'A new wing has been added to the school buildings ... The upper part is devoted to a chemical laboratory and lecture room on a splendid scale and convenient in every way' (Anon., 1887c: 169).

Figure 6.7: The chemistry lecture theatre at Manchester High School for Girls

Source: Archives, Manchester HSG

Chemistry laboratories of the 1890s to 1910s

Up to now we have provided accounts of ordinary rooms – usually attics or basements – that were converted into chemistry laboratories. Though at a very few girls' schools, such as NLCS and MHSG, specialized chemistry laboratories were built in the 1880s, the first widespread wave of construction of custom-built chemistry laboratories came in the 1890s and 1900s. For example, at Belvedere School, the new chemistry laboratory was opened in 1905 (Cowe, 1905). Alison MacKenzie recalled the excitement in the *Belvedere School Liverpool Chronicle*:

> By the time I was old enough to be trusted with a test tube and a bit of magnesium ribbon, which I dearly loved, the new Lab. over the gymnasium had been built. We were vastly thrilled when the workmen's dust-sheets were removed and a new doorway disclosed, which led through to these two new Labs. I well recall the momentous day when I was appointed laboratory assistant to help each morning to put out the apparatus for the classes. Alas, I wasn't much of a scientist in those days and must have been a sore trial.

> (MacKenzie, 1930: 46)

Earlier in this chapter we noted that the first chemistry laboratory at Roedean had been an attic room. On moving the school to a new site in the late-1890s, a very spacious suite of science facilities was constructed causing great excitement among the chemistry students (Waller, 1899: 84–5). In addition to a large chemistry laboratory with a balance room, there was a science lecture room and a library that was also used by the science mistress for her own experiments.

Laboratories of the 1920s to 1930s

During the inter-war period, many schools built new large, spacious, and well-equipped laboratories. For the schools we visited, the earliest to build at this time was at Croydon High School in 1919: 'The Governors of the School have now provided excellent accommodation for Science work by building an extension to the old Chemistry room …' (Anon., 1924: 16). The next year, 1920, Ipswich High School installed a new laboratory as was described in the *Ipswich High School Magazine*: 'A new Chemistry Laboratory has been built next to the existing laboratories, an addition rendered necessary by the size of the school and the requirements of the Advanced Course' (Anon., 1920: 2).

The design of the new Sutton High School laboratory (Figure 6.8) was a collaboration between Miss Hunt, chemistry teacher at Sutton (see online Appendix 2), and Miss Lees (also online Appendix 2), who had retired from teaching chemistry at Clapham High School: 'The building programme was now to begin. From 1925 to 1928 it was the laboratories, planned jointly by Miss Lees and Miss Hunt, and we know that in their time they were among the best in English girls' schools' (Anon., 1964: 64).

Figure 6.8: Annette Hunt (far right) in the Sutton High School chemistry laboratory, c. 1900

Source: Archives, Sutton HS

As in the three examples above, reports of the provision of new laboratory facilities in the inter-war period were mere brief notes, in contrast to the detailed descriptions of the labs of the 1880s to 1910s. Chemistry laboratories were largely taken for granted. An exception was Belvedere School, where two students, Barbara Gollin and Lois Buckley, enthusiastically reported in 1924 on the new laboratory [succeeding the 1905 laboratory described earlier] in the *Belvedere School Liverpool Chronicle*:

> We have had to wait patiently for a long time for the new lab., but now that it has materialized it has surpassed our wildest expectations. When the Modern Studies people beheld its shining glories, then were they truly grieved that they had forsaken the enlightened cause of Science for the sombre path of Modern Studies.

> (Gollin and Buckley, 1924: 33–4)

During the 1920s and early 1930s at some of the schools, not just chemistry laboratories but entire new science wings or buildings were constructed. As examples, a science block at Bedford High School was completed in 1927 (Godber and Hutchins, 1982: 103). The new chemistry laboratory is shown in Figure 6.9.

Figure 6.9: Bedford High School chemistry laboratory, possibly 1930s
Source: Archives, Bedford HS

At Redland High School it was the fiftieth anniversary of the founding of the school that coincided with, or perhaps caused, the completion of a science building as was described in the *Redland High School Magazine*: 'Though the inside [of the Sciences Building] was still unfinished [in June] the Chemistry Laboratory was in use, fitted temporarily with the old benches from the Lower Lab' (Taylor, 1932). At Roedean School, a whole new science building, with an up-to-date chemistry laboratory (Figure 6.10) was completed in 1928 (Anon., 1928).

Figure 6.10: Roedean School chemistry laboratory, 1930s

Source: Archives, Roedean School

Examiners' reports

We have shown so far that chemistry was taught at the girls' schools, but of what quality? To establish credibility external examiners visited the schools to ensure the girls met the proper academic standards.

At several schools, one or more reports of the examiners over the years have survived. For example, at MHSG, the *School Reports* (Archives, Manchester High School) of the 1880s and 1890s commented on the results of chemistry theory and practical examinations. The 1885 *School Report* noted that, for the chemistry examination, students were expected to have studied Miller's *Organic Chemistry* (Miller, 1857) and the introductory chapter of Tilden's *Chemical Philosophy* (Tilden, 1884). In 1890, for the senior girls, it was remarked that 'practical characteristics of basic and acidic oxides should be more carefully studied' while it was acknowledged that the junior girls had good practical knowledge' but needed to give more attention to an accurate description of the 'results actually obtained' (Anon., 1890: n.pag.).

The examiners' reports at Mary Datchelor School were summarized by the headmistress, Florence Grimshaw, and included in the subsequent issue of the *Datchelor School Magazine*. For example, in 1900, the Cambridge Syndicate Examiners stated:

> Forms VI and Upper V were examined by papers restricted to *Chemistry* only. Here more attention should be paid to the experimental side of the science. There was a thorough knowledge of fundamental principles, laws, and facts, but some weakness was shown in applying this knowledge to the devising of experimental proofs or illustrations. Still, the work was very well done and some of it was excellent.
>
> (Grimshaw, 1900: 16)

For the GPDSC schools, a complete set of reports exist specifically on chemistry from 1884 through until the end of our period of study. These reports involved visits by examiners to all the GPDSC schools and the examiner was, at times, scathing in their criticism of a particular school.

In 1884, the examiner, W.W. Fisher of Corpus Christi College, Oxford, visited seven GPDSC schools, of which only Croydon High School gained his praise. He gave some overall impressions of chemistry at the GPDSC schools, including:

> Most of the girls answered correctly questions about the gases entering into the composition of the air, and gave methods for preparing common gases, such as oxygen, hydrogen, carbonic acid, and the like; a majority of them successfully worked the problem set. But it appears evident from the answers, that the girls have little knowledge of the things they describe and write about – and are only reproducing book-work. The teaching does not appear to have been sufficiently experimental, or illustrated by actual experiments.
>
> (Anon., 1884b: 73–4)

By 1886, an anonymous examiner found a great improvement at all the schools and he was impressed by the chemistry examination results from all but 'one or two' of the GPDSC schools. He then discussed the results for individual schools, Gateshead High School being particularly praised:

Division I. This Form took a paper on the South Kensington Advanced Course [in Chemistry]. Of the nine girls two were good, two moderately good, and the rest very weak. The same Form took a paper in Practical Chemistry. The work was, as a whole, extremely good, only two or three girls failing to find all the bases and acids set. The chief deficiency was in drawing conclusions from experiments, the results of which seemed in many cases scarcely warrant the conclusion arrived at.

(Anon., 1886: 150)

The construction of chemistry laboratory facilities – at first very basic, then later much more sophisticated – gave girls the opportunity to do practical chemistry, and they embraced it with glee as a new, exciting experience. In the next chapter, we show that the enthusiasm resulted in the formation of science clubs, many of which had chemistry-related activities as their focus.

Chapter 7

Chemistry and school science clubs

A minor, but nevertheless interesting, item of the varied programme
[of demonstrations by the science club] was the electrolysis of
"Aqua pura," vulgarly known as H_2O, and the decomposition of
salts by electrolysis, using antediluvian apparatus.

('M.E.', 1922: 16)

The enthusiasm of many of the girls for science – particularly for chemistry – can be found in their interest in science clubs and societies. Our source of information on the science clubs and societies was the student school magazines. Particularly in the early decades the magazines of the respective schools contained lengthy accounts of their club and society activities. We had access to the early magazines of 59 independent girls' schools, of which about three-quarters had active science clubs or societies at some time during our time frame, while four others had science activities without a formal club.

Founding of science clubs and societies

The goals of the science club or society (or field club, as some were originally called) differed between schools. However, each club was active in one or more of the following areas:

- organizing meetings for the discussion of scientific topics among the students, sometimes with demonstrations
- organizing meetings with visiting speakers, sometimes with demonstrations
- organizing expeditions, particularly to industrial plants
- organizing annual exhibitions, either for the school community and/ or for parents
- producing a science magazine and/or instituting a science library.

In terms of chronology, of the schools visited there was a group of clubs formed prior to the First World War and a group in the inter-war years.

Early science clubs and societies

In chapter 2 we noted a report on the formation of a science club at Polam Hall in 1907. By contrast the Mount School had a flourishing scientific society during the 1920s. Then, in chapter 3, we described the very active science club at North London Collegiate School (NLCS), which was founded in 1890 and still remains active. Also, we mentioned that Cheltenham Ladies' College (CLC) had a field club dating from 1889.

Figure 7.1: An early expedition, Cheltenham Ladies' College, 1901
Source: Archives, CLC

Among the schools we visited the earliest science club of all was the Science Circle of Wimbledon High School. Founded in 1889 (Anon., 1889), Martha Whiteley (see online Appendix 3) was reported in the *Wimbledon High School Magazine* as having given a lecture on the spectroscope in April 1892. However, only five attended the meeting and so it was decided to dissolve the group (Anon., 1892: 5–6). An early field club was that at Worcester High School (later renamed Alice Ottley School). It lasted longer than the Science Circle of Wimbledon, being founded in 1890 (Anon., 1890: 37) and ceasing in 1904 (Anon., 1904a: 1064).

Wimbledon House School (later renamed Roedean School), was also a pioneer, with a natural science club founded in 1893 with much enthusiasm. According to the *Wimbledon House School News*: 'The first meeting was also, as it were, a kind of house-warming for the new Laboratory ... There are between 30 and 40 members of this Club and the attendance on all occasions has been admirable' (Anon., 1893: 11).

Founded in 1896, the science society of Winchester High School (later renamed St Swithun's School) was particularly active. The strict rules laid down for membership were spelled out in the *Winchester High School Chronicle*:

> Any Member of the School can become an Ordinary Member by giving in her name to the Secretary, and shall remain a Member as long as she remains at the School, unless she request to have her name removed from the list of Members or fail to fulfil the condition of Membership specified below, namely: – No one shall remain a Member unless she shall have herself collected and exhibited at each Annual Exhibition, subsequent to her date of joining the Society ...
>
> (Anon., 1896–7: 15)

There was a real enthusiasm for science at many of the schools, as evidenced by the formation of the field club (later science club) at Colston's Girls' School in 1903: 'The study of Science which is extremely popular in our School has become even more attractive by the formation of a Field Club ...' (Anon., 1903: 5).

The purpose of the club or society differed slightly from school to school. The scientific society of Exeter High School (later renamed the Maynard School) held its inaugural meeting in 1910. According to the report by the secretary, the purpose of their society was to increase interest in science:

> Its object was to increase in the School the interest in many branches of science. It was arranged that about three lectures should be given every term on some scientific subject, and that the members only should attend the meetings. The number of members at the beginning was seventy-eight and has remained at about that number.
>
> (Roper, 1910: 9)

Inter-war science clubs

As mentioned above there were some science clubs and societies founded in the early years, mostly during the 1890s and 1900s. Then there was a second burst of enthusiasm for science societies in the inter-war period.

At Bedford High School for Girls, they called their society members the 'Ionians'. The name derived from the term 'Ionian Enlightenment' which was the appellation for the set of advances in scientific thought centred around Ionia of ancient Greece in the sixth century BC (Freely, 2012). The Ionian Society at Bedford flourished through the 1930s, with a report in *The Aquila* in 1937 stating: 'This year, there are more Ionians than there have been in previous years, and our membership numbers well over one hundred' (Anon., 1937: 10).

At Walthamstow School for Girls, the science society report of 1920 in the *IRIS* makes it clear that scientific outings were clearly the major focus:

> The Science Society was the first Society to be formed in the School. It originated in 1916 at the request of several girls in L.IVA. ... We have now about ninety members, most of whom are really keen on this part [scientific expeditions] of our work.
>
> (Jupp and Smith, 1920: 8)

The numbers of girls interested in science at many schools has to be considered as astounding. And it was not just limited to a few of the schools, but seems to have been a widespread phenomenon. At Wycombe Abbey School the science club had been founded in 1918 (Anon., 1918: 116) and by 1933 it was reported that: 'Numbers in the Science Club have increased greatly this term and now number over 100' (Anon., 1933: n.pag.).

Figure 7.2: Wycombe Abbey chemistry laboratory, 1911
Source: Archives, Wycombe Abbey School

The numbers were so large at some schools that a separate science club for the juniors was started. As an example a junior science club was organized at Belvedere School in 1923: 'In the Summer Term a Junior Science Club was formed and a few of our younger members in IVa, left us to join it' (Buckley, 1923: 34). At Sutton High School there was a rigorous admission standard even for the junior branch of their science club. A club report in the *Sutton High School Magazine* noted:

> A Junior Branch of the Science Club has been formed for girls in the Lower IV and Upper III Forms. In order to qualify for membership, these girls must pass a short test on general scientific knowledge or demonstrate before a committee their interest in some scientific subject. The Club has, at present, a membership of thirty.
>
> (E.M.A., 1933: 30)

Student chemistry presentations

The clubs catered to all the sciences so there was a wide range of topics among the student presentations. Nevertheless, a significant number of presentations were on chemistry and some of these were illustrated with demonstrations.

History of chemistry presentations

The history of chemistry, especially a discovery that could be interwoven with the life of a famous chemist, was always a popular topic at science club or society meetings. For example, a student member of the science society of Malvern Girls' College spoke on the discovery of the element oxygen:

> Hilda [Donald] gave a paper on the "Discovery of Oxygen." There are three names associated with the discovery of oxygen – Scheele, Priestley, and Lavoisier. They made experiments with air and discovered that one part supported combustion while the other did not ... Oxygen is essential to the support of animal life and there is only one element (fluorine) into combination with which it does not enter. Two common methods of preparing oxygen are (1) heating Barium Peroscide [sic: Peroxide], and (2) heating Potassium Chlorate with Manganese Dioxide.
>
> (Anon., 1916: 17)

Figure 7.3: Malvern Girls' College chemistry laboratory, c. 1920
Source: Archives, Malvern Girls' College

Alchemy was another common topic, as this account in the *Winchester High School Chronicle* from the science society relates:

> Molly Cook read a paper on 'Alchemy'; she explained the theories of the alchemists, and showed how the study of alchemy differs from that of modern chemistry. She then described, with the aid of excellent drawings done by Faith Crickmay, some of the apparatus used by the alchemists. At the close of the paper, Miss Verinder gave an account of the properties of radium, and showed that the alchemists were not so far from the truth as people think.
>
> ('F.M.R.', 1918–19: 16)

Presentations on contemporary topics

Though its compounds were first discovered in 1898, the highly radioactive chemical element radium was not actually isolated in its elemental metallic form until 1910. As a result its properties became a source of fascination for students of many science clubs and societies, such as those of the scientific society at Exeter High School: '... several of the girls read papers on Radium, explaining its chemical and medical uses' (Anon., 1911: 9).

Contemporary events were the focus of other presentations. For example, in the latter part of the First World War, some talks related to the war effort, including this paper presented at the science society of Winchester High School:

In the Autumn Term, T. Rose read a most interesting paper on "Some Aspects of Chemistry During the Present War." She described the structure of a typical shell, giving illustrations on the blackboard, and then discussed the nature and properties of some of the high explosives employed in the manufacture of the shells. She also referred to the important part played by dye works during the war, explaining that they were the only works fully equipped for the manufacture of high explosives at its outbreak.

('J.O.E.', 1917–18: 15)

Unfortunately, many of the accounts are very brief and the experiments are not described in any detail. For example, at the Natural Science Club of St Paul's Girls' School, it was reported that: 'Dorothy Bloss read a paper on "Pyrotechny," a subject of great interest to the audience. It was illustrated by experiments performed by the Secretary and Treasurer' (Anon., 1925a: 6). At a senior science club meeting at Belvedere School it was noted that: 'Two Papers were read, the first on "Nitrogen," by A. Prestwich, and the second by C. Chorlton on "X-Rays." Both were illustrated by experiments and were most interesting' (Buckley, 1923: 34).

In the first half of the twentieth century coal was both a source of energy and of chemical raw materials. The topics were addressed in several presentations. For example, at Streatham Hill High School, coal products were described and shown: 'Lizzie Evans read a paper on "Coal and some of its Products". Several specimens of aniline dyes and other interesting and numerous products were on view' (Anon., 1928a: 17). And the following year, the chemistry of coal combustion and its use in gas works was illustrated by experiment:

A meeting of the Science Club was held in the Chemical Laboratory, ... when Adèle Perrick read a paper on "Combustion." This was illustrated by diagrams, and at the conclusion of the paper experiments were carried out demonstrating the uses to which combustion is adapted in modern industry. A model of a gas works had been set up and coal-gas was collected from it and tested in various ways.

(Anon., 1930: 21)

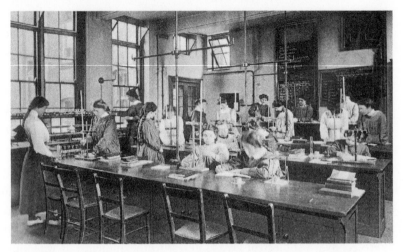

Figure 7.4: Streatham Hill High School chemistry laboratory, undated
Source: Archives, Streatham HS

Demonstrations by students

In some cases we do have details of experiments performed, such as this early account in the *The Paulina* by the field club of St Paul's Girls' School:

> K. Grant then performed some very effective colour changes, after which M. Hooke showed the surprising results obtained by burning various substances in oxygen and in chlorine. An amusing diversion was created when too much sulphur dioxide was evolved, with the result that many of our guests were almost asphyxiated. After strenuous efforts to extinguish the flames, the Vice-President [of the Club] gallantly bore off the obnoxious liquid.
>
> (Seaton, 1909: 8)

The properties of liquid air held a particular fascination. Experimentation was usually undertaken by teachers, visiting lecturers, or at industrial plants, but at Streatham Hill High School the girls themselves demonstrated the properties to the science club, even the highly dangerous reaction of burning sulphur with liquid oxygen:

> As the subject of the paper promised to be very interesting, there were a hundred and four people present. Doris Mathews and Betty Boyd read a most interesting paper on "Liquid Air," which they supplemented with exciting experiments. Grapes, bananas, oranges, and tomatoes could only be broken with a

hammer after immersion in liquid air. A rubber ball cracked like an egg; a frozen piece of beef was hard enough to crack a plate. Mercury made an excellent hammer with which nails were knocked into a board. The last experiment was perhaps the best. A hole had been made in a large block of ice, into which liquid oxygen was poured. A piece of burning sulphur was then dropped in, and it burnt with a glorious mauve flame that lit up the whole room. The meeting ended with a hearty vote of thanks to Doris and Betty.

(Anon., 1928a: 16–17)

Figure 7.5: St Paul's Girls' School chemistry laboratory, c. 1930
Source: Archives, St Paul's GS

In 1922, the *Downe House Magazine* carried a lyrical account of a demonstration of the electrolysis of dilute sulphuric acid by two members of the science club:

I do not think the school sufficiently appreciates its Science Club. Does it realize that two members of that unrivalled society have performed a hitherto unachieved feat of ingenuity? I think not! ... In a calm, clear, concise way, in a spacious lab, with suitable apparatus, they demonstrated to us ... There before our admiring eyes, water, that pure liquid we knew so well and used so often, was split up. "Split up!" you say incredulously; and yet it was so – no longer did we know it, for it decomposed into oxygen and hydrogen, or rather hydrogen and oxygen, as the latter was loth to enter into its appointed test tube! With what breathless silence we watched the famous scientists dip their hands callously into sulphuric acid, and at last, when called to our evening studies, how sadly we tore ourselves away from the scene of triumph, where the ingenuity and intelligence of the Science Club had been so well vindicated.

(Anon., 1922: 17)

Professional speakers

It is of particular note that career opportunities related to chemistry were sometimes among the lectures given. So there was an implicit assumption that chemistry was not being taught simply 'to be as good as the boys' but with the possible intent of using it as a pathway into career opportunities. For example, at Wimbledon High School, it was reported that: 'Miss Iva Lewis [see online Appendix 2] read a paper on "Chemistry as a Career for Women"' (Eborall and Kaye, 1924: 32). And at Roedean, a lecture was given by Ida Smedley Maclean (see online Appendix 3) of the Lister Institute. The report of the event noted: 'Mrs McLean [sic] showed the wonderful possibilities of *"Applied Chemistry as a Career for Girls"*' (Anon., 1919: 48).

Lectures with experiments

Lectures that involved experiments were a particular highlight of the science club programmes. The science club at Downe House enjoyed a lecture, with experiments, on plant chemistry:

> Dr Druce, MSc, of Battersea [John F. G. Druce, chemistry master at Battersea Grammar School] came down and gave us a lecture in the new lab on the "Chemistry of Plants." ... He did several very interesting experiments such as the extraction of chlorophyll from leaves and proving the presence of many of the common metals in the ash of hay.

(Anon., 1924a: n.pag.)

Figure 7.6: The new chemistry laboratory at Downe House, Berkshire, 1924

Source: Archives, Downe House

It was crystals that enthralled at Wimbledon High School:

> Dr Rosenhain [Dr Walter Rosenhain, Superintendent of the Metallurgy Department of the National Physical Laboratory] very kindly gave us a lecture on "Crystals", illustrated with lantern slides, specimens, and experiments. Since this lecture several crystals have been growing at school with great success.
>
> (Shears, 1928: 34)

For some schools, particularly in the greater London area, science students were taken to lectures at other institutions. As an example, the senior science girls at Streatham Hill High School were taken to lectures and demonstrations on physical chemistry:

> Miss Dalston [see online Appendix 2] and Miss Way took the upper sixth science to some interesting lectures on physical chemistry at the West Norwood Institute. They were illustrated by many experiments performed by the students in the laboratories upstairs. These lectures proved both instructive and helpful.
>
> (Anon., 1935: 10)

Several school parties attended lectures at the Royal Institution. In 1929, for instance, students at Blackheath High School were taken to three lectures there:

During the spring term some members of the Form VI were taken to three lectures at the Royal Institution, one given by Sir William Bragg on "Crystal Analysis," the second on "The Distribution of Chemical Elements" by Professor Goldschmidt, and the third on "Penetrating Radiations" by Sir Ernest Rutherford.

(Anon., 1929: 6)

Radium demonstrations

As mentioned above, the discovery of radium and radioactivity was a major source of excitement during the 1910s and 1920s. In addition to the girls themselves speaking on the subject, many science clubs had visiting lecturers on the topic. The inherent dangers of radioactivity were not appreciated and samples of radium compounds were widely available and handled openly. At Worcester High School the lecturer was a visiting university professor: 'A lecture on radium was given by Professor Poynting of Birmingham University [John Henry Poynting, professor of physics]. The lecture was splendidly illustrated with experiments and lantern slides' (Anon., 1904b: 1064). While at Downe House it was a science teacher who performed the experiments with radium: 'Miss Phelips gave a display of practical experiments including the decomposition of water by radium' (Anon., 1925b: 14).

A party from the City of London School for Girls was taken to King's College, London, to view a lecture with demonstrations:

… we attended the lecture on Radium and X-rays … The lecturer performed the following demonstration to illustrate this [the rays from radium causing other substances to become luminous]. He had a large flask the inside of which was coated in zinc sulphide. A closed glass tube was fitted into the cork of the flask which contained a very small quantity of radium emanation [radioactive radon gas]. The tube was opened, thus allowing the small quantity of the gas to pass into the flask, and immediately the bulb of the flask was filled with a green glow due to the fact that the small amount of radium emanation had caused the zinc sulphide to become luminous.

(Anon., 1924b: 80)

Liquid air demonstrations

The properties of liquid air were a highlight of meetings at many of the schools. Though at Streatham Hill High School, the girls had performed the experiments themselves, at all other schools it was a science teacher, or a visiting speaker, or the experiment was performed during a visit to an academic institution. One of the more eloquent accounts was given in the *St Swithun's School Chronicle* of a lecture by Dr P.T. Freeman of Cambridge University 'accompanied by a most exciting demonstration':

> Flowers dipped into it [liquid air] fell to pieces at a touch, the water in the cells of the petals having frozen by the intense cold; grapes were changed to glassy balls that could be smashed into fragments by a hammer, and the fact that raw meat was instantly frozen to the same glassy hardness was an unpleasant reminder of the inevitable fate of any hand that should inadvertently come into contact with this harmless looking liquid. That, on standing, liquefied air tends to become increasingly pure liquid oxygen was shown by the fact that felt and other materials, after being dipped in it, burned brilliantly and with almost explosive violence.
>
> (Woodard, 1932–3: 19)

Freezing liquid mercury was also a popular part of liquid air demonstrations, as was the burning of sulphur in liquid oxygen. However, not all of the reports explained, as the one above did, that it was liquid oxygen that caused the spectacular combustions.

The following series of demonstrations at St Martin-in-the-Fields School illustrates how liquid mercury was handled as an everyday substance. The hazard of burning sulphur in pure liquid oxygen and the toxicity of sulphur dioxide gas were not considered either. Instead, the science club secretary, Marguerite Dubois, described the 'exciting experiments':

> At our next meeting during the Spring Term, Miss Sutton [science teacher – see online Appendix 2] gave a lecture on "Liquid Air," and did many exciting experiments with it. After she had shown us a kettle containing liquid air boiling on ice some of us were inclined to be sceptical and think that liquid air must be very hot, but when we saw her bang nails into a block of wood, with a hammer made by freezing mercury in liquid air, we were convinced to the contrary. A very beautiful effect was obtained by dropping a piece of burning sulphur into a little pool of liquid air in a hole scooped out of an ice-block. The sulphur burned with a lovely blue flame, which appeared very beautiful when seen through the ice.
>
> (Dubois, 1932: 43)

Expeditions to industrial plants

The science students found chemistry fascinating. But when it came to visits to industrial plants, as the reader will see below, such words as 'thrilling' permeate the accounts. To that generation of girls, industrial chemistry was pure excitement. However, the authors of the reports were always careful to include the chemical aspects of their visit. The girls seemed fearless, for some of the accounts described red-hot metal, flames, toxic gases, and so on (as had the NLCS students experienced, as described in chapter 3).

Visits to gas works

Without exception every science club or society visited at least one gas works. Urban Britain relied heavily on coal gas for heating, cooking, and – in the early decades – for lighting. The hydrocarbon gas mixture was produced by heating coal in the absence of air (a process known as pyrolysis). Thus during our period of interest the gas works was the one chemical industry that every large community in England possessed.

These visits were not just to see the awe-inspiring buildings and processes but to understand the chemistry involved. The most descriptive account was given by Molly Baddeley of a visit by the science club members at Belvedere School to the Liverpool Gas Company's works at Garston:

> After walking a few paces we turned a corner and saw, in the near distance, the thrilling sight of a huge mass of burning cinders shooting down the wall of a large building, while from the bottom of this wall banks of smoke and steam arose. Our excitement was at once aroused, and we pressed eagerly forward toward this scene of splendour ... The series of furnaces is surrounded by a passage. We entered this passage and found ourselves in what might almost have been part of the infernal regions, for it was intensely hot and dark except for one spot, where, from an opening in the wall, huge flames were belching forth ...
>
> In order to show us the intense heat of the furnaces a small plug was taken from the wall of one. We gazed in awe at the almost white-hot interior, until the heat from the opening drove us away ... [Later] we saw the tests for sulphuretted hydrogen being made, by holding over the gas a paper dipped in a solution of lead acetate, which any sulphuretted hydrogen quickly turns brown. We realized, from the care taken in these tests, why all

our efforts to find traces of sulphuretted hydrogen in our gas supply have been in vain.

(Baddeley, 1925: 56–7)

Figure 7.7: Garston Gas Works, 1920s

Source: www.garstonhistoricalsociety.org.uk/
garston_industrial_development.html

Visits to sulphuric acid plants

In that era there were factories in several locations in Britain producing sulphuric acid. For nearby girls' schools, tours of these plants were also a favourite. The science club of Ipswich High School visited Messrs. Packard and Fison's works at Bramford, as recounted by Audrey Pipe:

> We then went through the furnace room, in which there was a strong smell of sulphur dioxide. In these furnaces copper pyrites are roasted with the formation of sulphur dioxide and impure copper. Each furnace is stoked at regular intervals, and we looked in several which were at different stages. Some were at red heat and the sulphur dioxide could be seen rising off the pyrites, some had given up all their sulphur and others had recently been stoked and were not yet red.
>
> (Pipe, 1929: 18–19)

A group from Clapham High School commented on inhaling the sulphur dioxide, and their tour seemed somewhat hazardous:

In November 1923, a party of the Upper and Lower VI Science girls was taken over the sulphuric acid factory at Greenwich. They were shown the pyrites burners where the sulphur dioxide is obtained, and the preparation of the oxides of nitrogen, used in the Chamber Process. They then passed on to the actual leaden chambers, and when later they mounted a ladder, and came on the roofs of the chambers, the sulphur dioxide was everywhere in their eyes and in their mouths. They also saw the finished sulphuric acid flowing out from the absorption towers.

(Anon., 1925c: 54)

According to the account in the *Laurel Leaves: Edgbaston High School Magazine*, ladder-climbing was also part of the experience for the chemistry students when they toured a combined sulphuric acid and hydrochloric acid plant:

We had been advised to take old mackintoshes, and were very glad of this as we had a lovely time climbing dirty ladders to view various parts of the workings ... We were very thrilled at being allowed to climb up a little ladder in pairs to look into the furnace in which the sodium hydrogen sulphate was being converted to sodium sulphate, which could then be sold.

('H.K.L. M.A.C.', 1937: 31–2)

Visits to other chemical factories

Soap manufacturers were another favourite industry for science club visits. Marjorie Davis of the science club described in *Colston's Girls' School Magazine*, their tour of Messrs. Christopher Thomas's soap works:

It was interesting to see all the stages in the process of soap-making from the beginning, when the fats were melted by steam, then pumped from the ground floor of the factory to the top floor, and there poured into vast cauldrons where, with water and caustic soda, they are gradually boiled into soap ... We also saw the vacuum chambers where the valuable by-product, glycerine, is distilled and purified.

(Davis, 1929: 37)

Located in north-west England, the Belvedere School was in an area rich in chemical industry. A tour of the United Alkali works at Widnes by the senior members of the science club was the highlight of 1921 as Lois Buckley described:

The party was conducted by four chemists, who explained the different chemical processes which were occurring in the various unfamiliar apparatus. We followed out, in detail, the manufacture of caustic soda from the raw salt. First, the brine was purified and was then run into a series of electrolytic cells. These cells were especially worthy of notice, as they are only utilized by this Company, and are not described in any text-book. After electrolysis had occurred, the solution within the cell consisted of about equal quantities of salt and caustic soda. This solution was then concentrated in a partial vacuum, the salt separated out, and the caustic soda was then isolated.

(Buckley, 1922: 75–6)

Though Sheffield High School did not have a science club, the girls of the senior science division were taken out annually during the late 1920s and early 1930s. In 1933 they toured the steel works at Templeborough, reporting:

We saw one of the furnaces being tapped, which proved a most thrilling sight. During this spectacular process the molten steel is run out of the furnace into a massive iron ladle, amidst showers of sparks; the heat was so terrific we had to stand some distance away in order to watch in comfort.

(Anon., 1933–4: 8)

Annual science exhibitions

Another common activity was an annual science show produced for the school and often for the parents as well. For example, at St Paul's Girls' School it was reported that the science club held a scientific exhibition for all the members of the school. They included the hazardous, mercury vapour-producing Pharaoh's serpent reaction in their repertoire: 'Among the attractions were the making of silver mirrors and Pharaoh's serpents in the Chemistry Room' (Anon., 1923a: 4).

Similarly, at King Edward VI School for Girls, Birmingham, two members of the science club organized a 'Chemical Magic' presentation. The Pharaoh's serpent reaction was obviously popular at the time as they, too, included it:

They succeeded, among other things, in "turning water into wine," producing a miniature snow-storm, and charming a beautiful serpent out of an ordinary crucible. Judging by the enquiries afterwards as to how they did it, this last feat seems to have been their greatest triumph.

(Forsyth, 1929: 29)

The scientific exhibition by the science club at Downe House was centred on the favourite of the time, liquid air:

> We were able to carry out many interesting experiments, such as the freezing of flowers, lumps of meat, and rubber balls, so that they broke like china when dropped on the ground; we were also able to freeze mercury. A ring had been cut in a block of wax, filled with mercury and the liquid air poured over it; the mercury froze into a solid ring which it was possible to hold up by a piece of string. The following experiments were also carried out: The action of sodium on water; ... copper plating; colour changes; crystal formation ... The school seemed interested and amused.

('S.L.', 1932: 35–6)

The science club of Wimbledon High School started an annual science exhibition in 1924, and these continued to the end of the 1920s. The report for 1929 described how their exhibition was serious science:

> At the end of the Summer Term, 1929, the Chemical, Physical and Botanical Laboratories were full of excited girls, dressed in black overalls, looking very worried but very learned. It was the Science Exhibition. I am sure every parent must have been most impressed and most interested by the various demonstrations ... Lower V showed the solubility of ammonia gas and hydrochloric acid gas in water. "Burning Air" (Upper IV), Invisible Inks (Upper IVB), "Supersaturation" (Lower V) and "Surface Tension" were other experiments which were much enjoyed.

(Stevenson, 1930: 33)

Some of the experimentation was quite advanced, such as that at Bedford High School performed by the students of the Ionian Society: 'One girl watched with zeal an intricate framework of test-tubes and apparatus extending the whole length of a work-bench. What was she doing? Not just distilling water or making some other simple experiment – she was preparing chloroform' (Anon., 1938: 8).

Science club magazines and science libraries

In chapter 3 we described how the students at NLCS had instituted a science magazine, some issues of which still survive. At least two other science clubs in London, those at Wimbledon High School and at Frances Holland School, also produced their own science magazine. Unfortunately no surviving copies could be found at either of the schools visited, so the only evidence of their existence is in science club reports.

At Wimbledon High School, there was an initial announcement in the school magazine: 'A [science club] magazine was brought out in the Summer Term, containing contributions from past and present members. It is hoped to bring out another number this summer ...' (Eborall and Kaye, 1924: 33). And a second mention was: 'For the second time a Science Club magazine was issued in the Autumn Term 1925, from the proceedings of the sale of which new books for the Science Library are being bought' (Anon., 1926: 28). In *The Graham Street School Magazine, Frances Holland School*, the sole mention of a science magazine was: 'It is proposed that the [science] Club shall have a Magazine, edited by H. Leach, which shall be purely scientific' (Anon., 1907: 12).

Several science clubs founded their own science libraries. At Clapham High School the chemistry teacher, Lilian Ida Quartly (see online Appendix 2), provided the science students with books. One of the old girls, Elizabeth Adams, recalled: 'Books appeared from Lewis's Library, ranging from Maurice Travers's "Discovery of the Rare Gases" to a translation of Lucretius's "De Rerum Natura"' (Adams, 1960: 2). 'Lewis's Library' referred to the Scientific Lending Library established by the bookseller H.K. Lewis of Gower Street, London (Boothby, 1964). Surviving into the latter half of the twentieth century, subscribers paid an annual fee (25 shillings in 1949) in order to borrow from their extensive collection.

The science club at Roedean School regularly added chemistry books to their science library over the period from 1913 to 1928. The titles and authors of these 21 chemistry books are listed in online Appendix 1.

The demise of the science clubs

The peak of interest in the science clubs occurred during the inter-war years. Whether the inter-war club flourished more in the 1920s or in the 1930s seems largely to relate to the enthusiasm of the chemistry teacher of the period. As an example, the science society of Notting Hill High School was active in the 1930s. It started with great promise in 1928: 'During the science society's first year it has shown itself both active and versatile in its pursuits' (Anon., 1928b: 9). But by 1939 it was no more: 'At the end of the summer term it was decided that the society should not continue unless the girls were keener' ('M.T.', 1939: 10).

The science club of Haberdashers' Aske's Girls' School, Acton, only survived part of the 1920s. A successor to the upper school nature study club, the science club was founded in 1921 (Swain, 1921: 8–9) but came to an end in 1925:

> At a meeting held this Autumn Term it was decided to dissolve the Club as the majority of the members, though good listeners, were not taking an active part in the work of the Club, probably in part from a lack of time, opportunity and knowledge. It is hoped, however, that it will be possible from time to time to have scientific lectures.
>
> (Anon., 1925d: 10)

Though the science circle of Wimbledon High School only existed from 1889–92 (see above), the inter-war science club survived for a longer period. The club started robustly in 1922 with 42 members (Anon., 1923b: 25) but ceased to exist in 1934, at the time when the number of students passing exams in chemistry decreased dramatically (see chapter 12).

At some schools science clubs faded away, then, years later, a new club was formed, without any recollection that there had been one previously. The students at Ipswich High School had founded a science club in 1907 that disappeared. Upon the arrival of a dynamic new science teacher, Miss Virgo (see online Appendix 2), in 1921 the school instituted a science club again: 'A Science Club for the girls of the Upper School was started in the Spring Term' (Virgo, 1921: 16).

A chemistry club had been founded at Streatham Hill High School in 1907 ('M.P.L.', 1907: 279). Then, in 1908, it was renamed the Science Club ('M.P.L.', 1908: 314). However, we could find no reports of club activities after the mid-1930s. At the now-combined Streatham and Clapham High School, a science club was proposed in 1942. The author of the report, Evelyn Clayton, seemed unaware of the long-lasting Streatham School predecessor:

> The first suggestions for a Science Club arose during the Autumn term, stimulated partly by the weekly play-readings organised by the Arts Sixth. It was felt that Science should not be regarded merely as an examination subject, as was the general tendency in schools, but as an essential part of everyday life. To encourage

this attitude to Science it was suggested that the older girls in the School should have an opportunity of discussing current scientific ideas and taking part in scientific activities whether they took Science as a school subject or not.

(Clayton, 1942: 10)

The existence of science clubs at independent girls' schools at some period of time between the 1880s and the 1930s seems to have been almost the norm. Many of these societies or clubs flourished with very large numbers of members (except where membership was limited). Chemistry demonstrations and outings to chemical plants – particularly gas works – was the highlight of many of the meetings. The accounts sometimes described happenings or adventures that to the girls were 'exciting' but which in today's safety-conscious environment would never be permitted. In the next chapter, we show that girls exhibited their enthusiasm for chemistry in a very different way – that of chemistry-related literary contributions to their respective school magazine.

Chapter 8

In their own words: Chemistry poetry and short stories

There's going to extremes,
And there's going at the seams,
And we all know that there is going too far;
But going to press,
I must confess,
Is the worst of any there are.

(Anon., 1897: 5)

In the late nineteenth and early twentieth centuries, the girls' school magazines contained reports of every happening in their respective school (Symonds, 1899). As we showed in chapter 7, accounts of science club or society meetings and expeditions were well documented. In addition, the magazines at most of the schools reserved space for short stories and rhymes/poems. We included in chapter 2 some of the contributions from The Mount School, and in chapter 3 from North London Collegiate School (NLCS) and from Cheltenham Ladies' College (CLC).

What amazed us was the widespread use of poetic verse and short stories as a vehicle for expression on the topic of chemistry. Such literary works were to be found in magazines from almost every one of the many girls' schools that we visited, both day school and boarding school. Such efforts seemed to be more common from each school's founding up until the late 1930s. In fact, chemistry-related literary works were not just a high school activity, but they were a form of expression used by young women at university, polytechnic, and also in the military workplace during the First World War (Rayner-Canham and Rayner-Canham, 2011).

Enthusiasm for chemistry

We begin this chapter with three prose accounts that demonstrate the excitement many of the girls felt for chemistry and experimental chemistry in particular. The first of these, from Downe House, was authored by 'M.G.' almost certainly Madge Godfrey, a student at Downe House from 1909 until 1912 (Kingsland, 2015). Reading her account, it was most definitely a different time from now: the author revelled in the thought of making dangerous gases and the even more 'thrilling' possibility of explosions:

The Lab

I choose this [topic] because chemistry is my favourite subject, and I can think of nothing better. We have shelves of most exciting glass apparatus, and we have got heaps of bottles, containing lovely coloured and exciting substances.

But the thrilling part is the experiments that you do in the lessons. When you have finished the rather dull course of "elementary physics," you begin chemistry. Then you analyse substances and find out all about them, and you are told quite cheerfully that today "we are going to make a gas that would make you insensible if you inhaled much of it!" But these little trifles add to the excitement, and we have never yet had a serious accident.

But perhaps the most thrilling things are explosions – we don't have many, but we have got as far as making water with an explosion of two gasses. Of course we have lots of minor experiments, but we make some extraordinary changes in colour and form of substances.

('M.G.', 1911: 19–20)

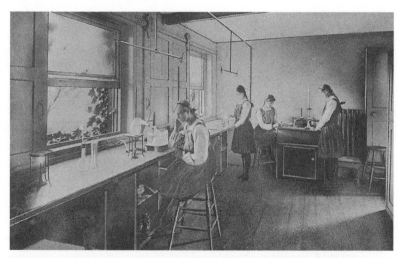

Figure 8.1: The chemistry laboratory at Downe House, Kent (formerly Charles Darwin's laboratory), photographed sometime between 1908 and 1915

Source: Archives, Downe House

'E.A.R.', at the City of London School for Girls, saw the chemistry laboratory as a place of potential disasters. Yet, despite all the vicissitudes, she still proclaims: 'Hurrah for chemistry!':

The Humours of the Chemistry Room

The chemistry student with bated breath, says: "Next spasm – practical chemistry! Oh!" in a voice that speaks volumes. She adjourns to the chemistry laboratory, dons her funereal garb, and forthwith proceeds to dissolve substance A. But, horror of horrors! Substance A is insoluble! Water, acids, alkalies, all are tried, but in vain. In an atmosphere compounded of acid fumes and "bad egg" gas, she wrestles with the difficulty, while to cheer her come the strains of "sweet music" from the practisers next door, and the fairy footsteps of the Badminton players upstairs. To add to her confusion she discovers that her bottle labelled "Ammonia" has no smell, and her concentrated acid is a culture-solution. Then the whisper goes round, "Stoppers in your bottles!" But to her dismay, she finds on her bench a small stopper without a bottle and a large bottle without a stopper. The book offers no help in such a predicament. But now substance A has been analysed with astonishing results. At least half a dozen metals are present. What a remarkable amount of impurity that simple substance contains!

('E.A.R.', 1909: 10–11)

The author of this third description of laboratory life attended the Central Foundation Girls' School, London (CFS). She reflects on the challenges of working with the equipment in the chemistry laboratory:

Chemistry

The benches are of a most uninteresting colour – a dull brown. Now, one day last term about eight girls, all with a love of things beautiful, baptized these benches with something mysterious from their store. This had the effect of changing them from dull brown to red and yellows of most glorious gipsy hues. Unfortunately these efforts to beautify the laboratory were not appreciated and instead of being thanked for making CFS brighter and more beautiful, we were actually told to report ourselves for spoiling the benches.

('A Chemistryer', 1903: 46–7)

Simple rhymes

The use of rhymes was widespread at independent girls' schools, very often using chemistry content.

Rhymes based upon common children's poems

In many examples a popular song or ditty was taken and re-worded to suit the topic. One example was the 1868 Irish children's song 'Ten Little Indians' (Opie and Opie, 1997: 333–4). Culleton (Culleton, 1995) showed that poetry was used by women munition workers in the First World War as a means of describing the hazardous life in the workplace. She also noted the common use of well-known nursery rhymes and poems, and that 'Ten Little Indians' was particularly used as a framework. In chapter 2 we included one such verse from The Mount School. Two other schools where girls reworked this rhyme as 'Ten Little Scientists' were St Paul's Girls' School in 1926 and Clapham High School in 1929:

> ...
> Seven little Scientists
> > Chemicals did mix
> One became gasphyxiate
> > So then there were six
> ...

> ('Globigerina', 1926: 11)

> ...
> Four little Lower Sixth doing Chemistry
> One blew herself up – then there were three
> ...

> ('Sixth at Home', 1929: 40–1)

Another commonly used framework for women's rhymes of the period (Rayner-Canham and Rayner-Canham, 2011) was Rudyard Kipling's popular poem *If* (Kipling, 1910). This next verse by a student at St Paul's Girls' School, adapted from the *If* format, mused about the possibility of her English teacher conjugating chemistry terms as verbs:

> WHY?
> Why can't I see Miss Webbert give a lesson
> > Of chemistry and conjugation mixed?

Why can't I say the passive of "to chlorine"?
  Or keep the gender apparatus fixed?

...

('Troubadour', 1925: 9)

Laboratory experiments in rhyme

Using a rhyming format, a student at St Paul's Girls' School described a whole laboratory experiment. The procedure involves the dehydration, then rehydration of blue copper (II) sulphate pentahydrate.

Experiments in Rhyme
Copper sulphate
Vitriol blue
Dissolve in water
To form solu.
Evaporate –
Water dries
Trichnic [Triclinic] crystals
Crystallize.
Heat the substance
Over gas
Result is white anhydrous mass.
Add a drop of H_2O
See the powdery
Whiteness go.
Back returns
The colour true,
Copper sulphate
Vitriol blue

('Bump & Bump Ltd', 1940: 14)

Tales of woe in the chemistry laboratory

Rhymes were also used to describe chemistry accidents, some fictitious, and some very real. The style of cheerfully cruel rhymes was that of Harry Graham (1874–1936) who authored the book *Ruthless Rhymes* in 1899 (Graham, 1899). In 1911, a small collection of Graham's ditties, recounting many sad fates of *Little Willie*, was published anonymously (Anon., 2011). The following quatrain to *Little Daisy* would seem to be very closely modelled on the *Little Willie* tales. It was the winner of a student competition for black humour verses at North London Collegiate School (NLCS) by M. Mackenzie:

Little Daisy

Little Daisy was always placid, –
Until one day she fell in the acid;
The mistress was sorry that she had gone west,
But said, "How nicely she effervesced!"

(Mackenzie, 1938: 510)

There was a second writer of such black humour poems: Hilaire Belloc. His *Cautionary Tales for Children: Designed for the admonition of children between the ages of eight and 14 years* appeared in 1907 (Belloc, 1907) and a second volume, *New Cautionary Tales: Verses*, was published in 1930 (Belloc, 1930), just prior to this next chemistry contribution. This series of light-hearted verses about the dangers of the chemistry laboratory came from K. Somers Cocks of Roedean School:

A Cautionary Tale

Peggy was a naughty child,
And at her lessons somewhat wild.
She found a test tube full of "water,"
And when small Elsie came and fought her
Poured it down the other's back
Till she was soaked, alack! alack!
But look at little Elsie now!
What has happened? What a row!
What ever happens in these cases?
Elsie's making funny faces!
Why! She's fizzing like ginger ale
And her face has gone quite pale.
Why! She's burning quite away.
Oh! Elsie, darling, with us stay.
But Elsie's growing less and less
And the floor's in such a mess
What a bang! Oh, Elsie's gone!
How bright the light was when it shone.
Peggy's crying now, you see:
"Now she's gone, and all through me."
The test tube was not full of water,
But the stuff called Pagnotater.
Never touch the bottles here,
But to chemistry go with care!!

(Cocks, 1934: 47–8)

Qualitative analysis and hydrogen sulphide

Qualitative analysis was a major aspect of school practical chemistry from the mid-1800s onwards. This activity involved the student being given a sample of unknown composition and being required to identify the ions present in the sample. The procedure owed its origins to the German chemist Carl Fresenius. It was in 1841 that he had devised a series of tests that could be used to stepwise classify and then identify which metal cations were present in an unknown sample. The procedure involved bubbling hydrogen sulphide gas through a solution of the compound or mixtures of compounds. For many of the cations a characteristically-coloured precipitate was formed.

Fresenius's analysis procedure was enthusiastically adopted as a standard method in British schools utilizing such monographs as: *Chemical Analysis for Schools and Science Classes: Qualitative Inorganic* (Scott-White, 1883). As a result of the use of hydrogen sulphide as a key reagent in the procedure, the 'rotten egg' odour became the immediately recognizable smell associated with a chemistry laboratory, though in times gone by its toxicity was rarely appreciated.

Composed by a student at Alice Ottley School, the verse at the beginning of chapter 6 included 'the smell of rotten egg' that came from the sulphide analysis procedure. This next rhyme by a student in the *Wimbledon Hill School Magazine* (WHSM) also referred to hydrogen sulphide. The structure follows the trochaic tetrameter of the *Song of Hiawatha* by Henry Wadsworth Longfellow:

To The Editor
Help my verses, come to bless them
For the WHSM
By the smell of putrefaction
On a Monday afternoon,
When you wonder what 'reaction'
Almost causes you to swoon,
Til a girl (with apron mystic
Spangled with an acid mess)
Tells you with an air sophistic
'We are making H_2S',
I adjure ye, come to bless them
For the WHSM

(Anon., 1890: 12)

The following set of nonsense verses from St Paul's Girls' School also refers to hydrogen sulphide, H_2S. Based on 'The Mad Gardener's Song' in *Sylvie and Bruno* by Lewis Carroll (Carroll, 1889: 78–90) the first line includes reference to a 'Kipp'. A Kipp (Sella, 2007) was a glass apparatus used to generate hydrogen sulphide gas whenever the valve was opened:

The Chemists Song
I thought I saw a Kipp last night
    'Twas making H_2S;
I looked again and saw it was
    A nut-meg playing chess.
If I had half-a-crown, I said,
    I'd buy a new Spring dress
        ...
I thought I saw an elephant
    Fall out of a burette;
I looked again and saw it was
    A brand new wireless set.
"Your hair is out of curl," I said,
    "You've lost your shingle net."

('Black and White and Red Seal', 1926: 12)

A chemical parody of the book *1066 and All That*

In 1930 a book was published entitled: *1066 and All That: A memorable history of England, comprising all the parts you can remember, including 103 Good Things, 5 Bad Kings and 2 Genuine Dates* (Sellar and Yeatman, 1930). It was a hilarious reworking of English history, with deliberate misspellings, for example, the medieval monk, the Venerable Bede, becomes 'The Venomous Bead'. Some students at Perse School for Girls adapted the idea to a chemistry context:

H_2O and All That
Chemistry is a memorable subject;
"Once smelt, never forgotten"
                        (Keatings)

Apparatus consist of Retorts (not to be confused with Backchat), Pinafores, Kipp-ers [used for generating hydrogen sulphide – see previous Section], Blast-Furnaces, Holmyard [author of many chemistry textbooks of the period], Farmyard, Hedges, etc. Most Chemistry, however, occurs in Test-tubes.

No chemical action is allowed to happen without an equation. A well-balanced equation is a thing of beauty and a joy for ever, and a sign of a scientific mind. Different equations happen in different atmospheres, others *must* have an atmosphere of work. Equations consist of Composition and Decomposition (Single, Double and Treble). Double Decomposition is the Paul Jones of Chemistry.

('The Science Study', 1934: 15)

Following the style of the English history textbooks of the period, the book, *1066 and All That*, also contained fictitious sample test papers interspersed among the chapters. These, too, were parodies, together with many unanswerable questions. For *H₂O and All That*, the students at Perse School for Girls similarly devised a mock chemistry test paper including such questions as:

Write the formula of potassium ferro-cobalti-nickelo-cupraammonium-methylethylpropylbutyl antimonyl tartrate.

Describe the preparation and properties of at least one of the compounds of Geranium. (HINT. Only found in *Rare Earths*.)

Are there any difficulties? If not, why not? (Give equations where possible.)

(ibid.: 15–16)

More sophisticated poetic works

Up to this point the rhymes have been light-hearted and often simplistic. Yet students at some other schools wrote serious poetic verse. The following poem on life in the chemistry laboratory, written by students at the City of London School for Girls, implies that periodic 'mild' explosions and fires were an accepted part of chemical experimentation:

A Ballad of Chemistry
On the bench we find strange compounds, which we must analyse,
But we look around in anger with very wrathful cries,
Our distilled water's disappeared;
We guess who thus our way has cheered,
Vowing vengeance, more we syphon
                        Utt'ring sighs.
We make fearsome weird decoctions and awful evil smells
The odour to the ignorant a tale of ruin tells.

Sometimes we soon get on the trail,
At other times we sadly fail,
But we all enjoy it greatly

Mids't our spells.

Sometimes we burn the benches, and at others singe our hair,
Or have a mild explosion, or quite a splendid flare.
You who sit and stew at German
And who read of noble Hermann
You can never know the pleasure

We find there.
(Anon., 1914: 29–30)

D. Francis, the author of this next contribution (published in *IRIS: Magazine of the Walthamstow School for Girls*), noted that she had used the poetic style of *I Have Been So Great a Lover* by Rupert Brooke (Brooke, 1971: 216–17). In Francis's chemistry adaption of the poem, she reflected upon her chemistry laboratory experience at the school:

The Laboratory

And I can cry with him, These things I love –
Bright-shining benches, beaming as they stand
With windows in their depths; and, arched above,
Taut question-marks of iron cold to the hand,
Their secret but half-kept, as from the gloom
One winking 'prisoned globule strives to fall,
Suspended quivering from its metal tomb.
I love the rows of shelves about the wall,
Glass-covered and be-labelled, where they bear
Thick-clustering bottles. Some are dark and brown
And blues look green through them, and scarlets wear
The curious shade of leaves that flutter down
Like birds with tired wings. But some are clear
As water in a beaker, and reveal
Warm crimsons, and the sober browns that peer
Through flaunting orange powders, to conceal
Their lack of colour. Then there is the shade
Of glowing copper – neither blue nor green,
And cow-slip yellow that can never fade.
Now some are crystals, with a glinting sheen,
And some dull powders, sanded, and opaque,
But all I love …

And lovely too are these:
Rainbows in prisms; the sound that burners make
When lighting, like the popping pods of peas;
Flat shining pools of liquid mercury;
The light and shade on flasks, or any glass
That's blown in curious shapes; the lazily
Curved swan-necks of retorts, where fluids pass
To writing veils of vapour; and, at last,
The little pangs of weariness that stab
When poison's under key, the locks are fast,
And darkness dims the lustre of the lab.

(Francis, 1938: 7)

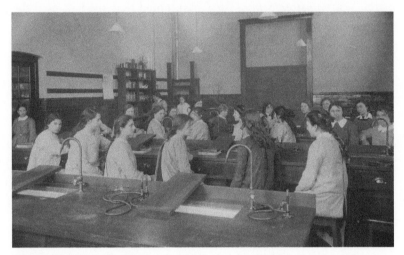

Figure 8.2: The chemistry laboratory at the Walthamstow High School for Girls, early 1900s

Source: Archives, Walthamstow SG

There was a similar expression of enthusiasm for the chemistry laboratory by K. Brown, a student at Nottingham Girls' High School. She referred to it as *Palace Beautiful* taken from the poem *The Palace Beautiful* by Orpheus C. Kerr, the pseudonym of Robert Henry Newell (Kerr, 1865: 11–21):

Lines to the "Lab"
O thou abode of scientific store!
How oft we dallied in thy precincts rare
Where none but those who delved in Science-lore
Knew half the mysteries that were hidden there!
The souls consumed with scientific fire

Within thy walls did feel there home aright;
To those who did not too high aims aspire
Thou wast indeed a palace of delight,
Where oft they spent so many hours of joy
E'en though they did not soar, and earned reproof
For some appliance used as 't' were a toy,
Or from busy workers kept aloof.
To some thou wast the half-way house to fame
To others "Palace Beautiful" thy name.

(Brown, 1900: 23)

Shakespearean parodies

The following poem uses the poetic style of 'All the world's a stage', a monologue from William Shakespeare's *As You Like It*. The chemistry adaptation in the *Bath High School Chronicle* was written by a New Girl after her first class in experimental chemistry. The student describes the heating of the element, sulphur, in a test tube. The solid element is a pale yellow powder. On melting its colour changes to brown. Some of the sulphur vaporizes and burns to give acid-smelling sulphur dioxide gas. Beginning at about 165°C (though the author of the poem states 157°C), the liquid sulphur becomes so viscous that it is almost solid. Then heating above 185°C causes the viscosity to diminish again:

A New Girl's Thoughts after her first Lesson in Practical Chemistry
(dedicated to Miss Davis [probably the chemistry teacher], with apologies to Shakespeare)
All the world's a test-tube,
And all the men and women merely Sulphur;
They have their properties and possibilities;
And one lump in its time plays many parts.
At first the yellow, gritty to the taste,
And then the dirty brown, with spots of black,
And muddy, bubbling face, hopping like frog
Within its tube of glass. And then the liquid,
Black as the River Styx, with filthy smell
Made to revolt its mistress' nose. Then treacle,
Full of great heat and sticky as the glue,
Brownish in colour, of adhesive quality,
Fixing itself on all things that it touches,

Even the thermometer. And then a substance,
In fair round test-tube with dark horrors lined,
With dreadful look and gloomier aspect yet,
Full temperature one hundred, fifty seven;
And so it plays its part. The sixth test shifts
Into the brown and nearly solid block,
With glass all round and tube poking about,
In its enormous heat, a world too great
For its small size; and its treaclyness
Turning again towards solidity,
Its colour still the same is always found.
Last stage of all that ends the Sulphur's tests,
Is second lumpiness, mere repetition,
Sans treacle, bubbles, smells, sans everything.

('P.B.', 1909: 16)

Figure 8.3: The chemistry laboratory, Bath High School for Girls, around the 1900s

Source: Archives, Royal HS, Bath

Students at the City of London School for Girls imagined William Shakespeare as the school inspector. Here are the lines relating to chemistry:

Chemistry

[Two separate groups of girls confront him here. One is standing around a Bunsen burner on which a water-pot is boiling.]

"Double, double, toil and trouble,
Fire burn and cauldron bubble"
(Macbeth)

[The other is gazing with intense interest at the demonstration of
the fact that a lit taper is extinguished when thrust into a jar of
carbon dioxide.]
"How ill this taper burns!"
(Julius Caesar)

(Anon., 1926: 43)

Reflections on the lives of laboratory coats

Reading of the acid spills and other assorted laboratory accidents, it is not
surprising that specialized outer clothing was required. As can be seen from
early photos, at some schools laboratory coats were more like smocks than
today's button-up style of coat. We found two accounts of the life of a lab
coat, the first a rhyme and the second a short story. The rhyme was by M.
Rodker in the *Central Foundation Girls' School Magazine*. It was closely
based on Marc Antony's oration in *Julius Caesar* (Act 3, Scene 2) by William
Shakespeare and here we have included the first half:

Oration on a Science Overall
If you have tears, prepare to shed them now.
You all do know this ov'rall: I remember
The first time that I ever put it on;
'Twas on a summer's morning in the lab.:
That day when I had donned my newest dress.
Look! in this place ran nitric acid through;
See what a rent the heated tong did make!
Through this the hydrochloric acid ran;
And, as I plucked my soiled ov'rall away,
Mark how the stream of acid followed it,
As though indeed it firmly were resolved
A strong impression on my dress to make!
For th' ov'rall, as you know, was my protector
Judge, O you girls, how dearly I did love it!
This was the most unkindest burn of all.
...

(Rodker, 1919: 30–1)

At the Oxford High School one of the students devised a daily diary in the life of a science apron used in the chemistry laboratory. This was a very lengthy contribution and two excerpts are given here:

Extracts from the Diary of a Science Apron

Wednesday, May 2nd. – I was taken up to a large room with my new friend, the Science Book, and flung over a green desk which smelt strongly of new paint. After a short interval my mistress sat on me stolidly for two solid hours. I feel quite faint still ... I was taken up, put on, and so entered a room with long tables and a curious odour, rather ancient and fish-like. I feel quite proud to think that this room is especially for *me* and my friend, the book. But I was horrified to find that a terrible burning liquid was smeared down my side. That girl should be more careful with her fingers!

Thursday. – Rested in the cloak room till the afternoon, when I was snatched up in a hurry and accompanied the Science Book to our room. I met a new smell today, it is *called* H_2S, but I think it is like ancient eggs. Several times I have been used for wiping dishes; I never expected to have such menial work. I shall protest.

('V', 1906: 413–14)

Figure 8.4: The Oxford High School chemistry laboratory, c. 1900
Source: Archives, Oxford HS

Poetry and prose related to chemistry teachers

At the schools we visited some of the poetic works in the school magazines revolved around the chemistry teacher. In the opening to chapter 5, we cited a rhyme dedicated to the retirement of the science mistress at Nottingham Girls' High School, Miss Ready. Here we include poems and a short story about three other science teachers.

Poems about Miss Cam

Though the poems about Marjory Thekla Cam were authored by students at Oxford High School, she only taught chemistry there for eight years, the majority of her career being spent at Bromley High School. Born in 1898 she was educated at Oxford High School and then at Girton College, Cambridge (1918-21). From 1921 Cam taught science courses at Oxford High School. During her time there two poems, printed in the *Oxford High School Magazine*, both with a certain amount of black humour, referred to Miss Cam. Then in 1929 Cam transferred to Bromley High School (another GPDSC school), where she stayed for the rest of her career from 1930 to 1946. She died in 1980.

The rhymes relate to fictitious accidents demonstrating the dangers of the chemistry laboratory. One was entitled: 'A Mishap in the Science Room' ('V2B', 1926: 1058–9), while the other is given below. In this second rhyme, by V. Wylie, the term 'nascent chlorine' is used. Chlorine gas was once believed to be a better bleaching agent if it was freshly formed in its 'nascent' (newly-born) state:

The Sad Fate of the Girl Who Meddled with the Science Experiment
In the playground could be seen
A belljar covering plants of green.
The jar was labelled "Do not touch,"
(Not that we wished to – very much.)
But one inquisitive small child,
Although abuse on her was piled,
Lifted that jar; alack! alas!
We smelt the smell of escaping gas!
Volumes of oxygen everywhere,
And chlorine (nascent) polluted the air.
Then these two gasses – ho! What a sight!
They bleached that girl to a deadly white.
She screamed, Miss Cam up at the sound,

And called the Fifth form all around,
Saying, "This is a good example of our
Lesson on chlorine's bleaching power!"

(Wylie, 1926: 1061–2)

Poems about Miss Lewis

Iva Gwendoline Lewis taught chemistry at Wimbledon High School for 30 years. She was born in 1891 and attended St Anne's, later named Abbots Bromley School for Girls, from 1902 until 1906. Then, after three years at Alexandra College, Dublin, she entered Trinity College, Dublin in 1910. Graduating with a first class honours BSc in experimental science (chemistry and physics) in 1914, Lewis taught at Roedean for two years (1915–17). From 1917 until 1921 she undertook war service in a government laboratory and factory from 1917 until 1921. All of her subsequent career (1921–51) was spent teaching at Wimbledon High School (Staff Records, Wimbledon High School Archives). Both poems described the unruly girls of the VIth form. One was entitled: 'Dedicated to the Science Staff' (Hazel, 1948), while the other is given below:

A Warning; or the Sad Story of the Worst Science VIth Miss Iva Ever Had

The chief defect of VI B Science
Was that, in absolute defiance
Of all Miss Iva's prayers and pleas
(She almost went down on her knees)
They never gave their note books in;
And, adding to this dreadful sin,
In spite of all her earnest pleadings
They never entered up their readings,
Which – mark, all scientists in the making –
Should be put down at time of taking.
They almost always tried to shirk
Preparing any reading work,
And, being peppered with abuse,
Would feebly murmur some excuse.
At practicals they were just terrors,
You've never seen such awful errors!
So huge these were, and so gigantic,
They nearly drove Miss Iva frantic.
Then dawned the day, alas, alack,

The last straw broke Miss Iva's back,
And she, rage following in its wake,
Put arsenic in their milk at break.

...

Next day, up to the lab, were found
Six prostrate bodies on the ground.
And when Miss Iva stood her trial
The judge said, with a kindly smile,
"We, after due deliberation,
Feel that you had much provocation.
I think their timely end well earned".
And so "Not Guilty" was returned.

(Loveless, 1942: 6)

A country ramble with Miss Stopard

The final contribution to this section is a short story recounting a country ramble by the science students at Sheffield High School with their teacher, Miss Stopard. Born in 1897, Winifred Mary Stopard was educated at Edgehill Girls' College, Bideford, and the Central High School for Girls, Manchester. She completed an honours BSc in chemistry at the University of Manchester in 1918. Stopard taught at Glossop Grammar School (1918–19); then the Central Secondary Girls' School, Sheffield (1920–3); and finally for the remainder of her career at Sheffield High School (1923–47) (Staff Records, Sheffield High School Archives).

We noted in the previous chapter that, even though Sheffield High School did not have a science club, the students were taken to a variety of chemical works in the 1920s and 1930s. Such expeditions were led by Miss Stopard. This account in the *Sheffield High School Magazine*, part of which is included below, cleverly incorporates a very wide range of chemical terminology:

The Scientific Side of a Country Walk

... We went bicar(b)onate acid drops on the way, until we got well out of town, when we decided to stop 'ard by the end of a lane. All progressed favourably until we came to a trough by the roadside, and one of our number had the misfortune to fall in! However, she was soon collected again by displacement of water, and quickly evaporated to dryness in the warm sun.

Figure 8.5: One of the illustrations accompanying the account of Miss Stopard's expedition.

Source: 'E.I.', 1931.

We then crossed a stile into a field, where flowers of sulphur and thistle funnels bloomed in great profusion. On somebody's remarking that she thought she heard ammonia, we discovered that the noise was coming from a dense cloud of bees evolving from a row of beehive shelves ...

By this time we were feeling rather tired, so we went towards a farm which we saw in the distance, in the hope of getting a glass of milk-of-lime. The farmer's wife was very busy shaking her (as)best(os) mats, but she gave us some saltcake, for which we had large apatites. She said her husband was suffering from a severe attack of plumbago, brought on by her having dropped, accidentally, an iron on his foot.

As it was getting late, and we had no wish to be left at nitride out in the country, we thanked the lady, and asked the way to the nearest main road. She said we would see her daughter Ethyl, who would direct us. We found the little girl with her friend Sal Ammoniac sitting on a bank, her hair tied back with a large bow of magnesium ribbon. They had bunches of rushes in their laps, and when we asked what they were doing, they replied that they were platinum ...

The girl, who was a very good conductor, led us to the road, where we were soon on our way back, and arrived ohm after a very enjoyable, and highly edifying day.

('E.I.', 1931: 248–9)

Short stories anthropomorphizing chemistry

There were some cleverly constructed short stories in which a chemical reaction was portrayed in human terms. One such example from Cheltenham Ladies' College, on the life of a carbon atom, was mentioned in chapter 3. U.R. Collingwood wrote a similar saga in the *Roedean School Magazine* (Collingwood, 1934: 48–9) that revolved around the carbon cycle and photosynthesis. What is particularly of note is the chemical knowledge shown in some of these anthropomorphized accounts. In a similar format, a story published in the *White and Blue: The Alice Ottley School magazine* ('J.B.', 1925: 1264) covered the essential chemistry of phosphorus.

From the *Orme Girls' School Magazine* comes this story by Margaret Tapley of which we include the first part. The saga commences with the storyteller, a nitrogen atom, being part of the compound, sodium nitrate, $NaNO_3$. The sodium nitrate is then converted by heating into sodium nitrite, $NaNO_2$, with the loss of oxygen gas. The sodium nitrite then reacts with ammonium chloride to give sodium chloride and ammonium nitrite. Warming a solution of ammonium nitrite produces nitrogen gas, N_2.

The Wanderings of a Nitrogen Atom

Crackle! crackle! crackle! What was that? My heart began to beat wildly – at last something was happening to me. For years, or so it seemed to me, I had lain in that tiny compartment with my companions, Mr Samuel Sodium and the three Masters Oxygen, without hearing any sound other than the continual chattering of my companions. Goodness! What could be happening to us? Mr Sodium sniffed audibly and remarked, 'Apparently our company does not suit the Oxygen brothers. See, they are leaving us.' 'No, only our eldest brother is going,' said the two younger Oxygens. And it was true – but how could we exist without him? – he had been our sole support for so long.

'Ssh! I see the Ammonium Chlorides approaching. Just watch Mr Sodium,' whispered the youngest Oxygen with an impish grin. I turned, just in time to see Mr Sodium shoot away from us, snatch Miss Chlorine from her companions and carry her away. I must say that neither the Oxygens nor I were sorry to lose Mr Sodium (the old grumbler) – but the Ammoniums were dreadfully upset. Suddenly someone detached himself from their group and came bouncing across to me.

'Whoever can this be?' I grumbled, but my tone altered when I saw who it was. It was a brother for me and at last I could go out into the world. For some time my brother and I lived peacefully in our new home, 'Atmosphere' ...

(Tapley, 1937: n.pag.)

An anthropomorphized story in the *Maynard School Magazine* uses the classic fairy story of the rescuing of a princess from a wicked uncle, who was a Baron. The students cleverly weaved the names of a large number of chemical elements and chemical terminology into a re-write of the tale. Only the first half is included here.

A Chemical Fairy Story

Once upon a time, in the far-away kingdom of *Europium*, there lived a beautiful princess in a *marble palladium*. Her name was *Beryllium*, and she had skin as white as *magnesia*, her hair was *lustrous*, her lips like a *strontium* flame, and she had large blueish-green eyes.

But she was very unhappy because she lived with her wicked uncle, Sir *Boron Litmus*, who was not of her *group* or *period* and so hated her. He appointed a *radical* governess for her, whose name was Miss *Acetate* and who made *Beryllium* do *qualitative analysis* from morn till night, till she yearned for *volumetric*.

But the princess had two faithful companion *constituents*, her dog, *Plutonium*, and her maid, *Ethyl Alcohol*, who always wore *starched* uniform.

In the little seaside *retort* of *Silicon-on-Bismuth* just near, there lived by himself a handsome prince whose name was *Oxygen*, and it was he whom Uncle *Boron* wanted to marry his daughter, *Aluminium*. But Prince *Oxygen* loved *Beryllium* best because she was not so cheap as *Aluminium*.

One day the princess vowed she would escape; so with the help of *Plutonium* she *filtered* through the great *ferrous* doors, in the middle of the afternoon, when her uncle and governess had forty winks in a *desiccator*, so they *krypt-on* to the Forest of *Argon* where the trees were *crystallizing* into bud and *methyl* oranges were hanging. She felt *positively electrolysed* by the liberation.

Suddenly she heard "*cobalt-cobalt-cobalt*"; it was the hoof-beats
and Prince *Oxygen* on his faithful steed, *Hydrogen*, appeared in
a cloud of *steam*, bringing for her a box of *Cadmium* chocolate
and a bunch of *flowers of sulphur* ...

('The Would-be Chemists', 1944–5: 24–5)

In previous chapters we provided students' accounts of wonderful new
chemistry laboratories and of thrilling outings to chemical plants. But such
discourses were within the student's 'world of chemistry'. Here we have
provided documentary evidence that the girls saw chemistry as an integral
part of their education and life. What seems particularly impressive to us is
the painstaking effort these contributors made to write an erudite work of
prose or to find an established work of poetry to re-craft into a chemistry
context. Beyond doubt, these girls had a passion for chemistry.

Chapter 9

Chemistry at some Welsh girls' schools

... anxious [auction] bidders quibbled over [cookery and chemistry] overalls, the most attractive of which had red and green patches in place of pockets, in addition to a yellow collar and a white belt, the ground colour, where visible, was a deep navy, splashed with black. This was rumoured to have belonged to one of the original members of the Brighter Chemistry for Girls Brigade – one, D-ph-e B-b-r.

('E-d R-b', 1930–1: 32)

Wales had a distinct history of education – especially for girls. In addition, there was no single, or even pair, of narratives, as was the case in England with Beale and Buss. Here we first endeavour to provide a brief and superficial summary of the development of academic girls' schools in the period 1858 to 1900. Then we focus on three girls' schools for which we could find some significant chemistry information: Howell's School, Llandaff; Cardiff High School for Girls; and the County School for Girls, Pontypool. Finally, we look at the Welsh branch of the Association of Women Science Teachers.

Girls' academic education in Wales

In England we can date the rise of the academic girls' school to the founding of North London Collegiate School (NLCS) by Buss in 1850 and the arrival of Beale at Cheltenham Ladies' College (CLC) in 1858. Similarly, we can identify the founding of the earliest academic girls' schools in Wales to the date of 1858 – though in the Welsh context they did not become role models.

The Howell's Schools

In 1540 Thomas Howell of Monmouthshire, a member of the Drapers' Company of the City of London, died leaving a significant sum of money in his will. These funds were to provide dowries to four 'maidens' and any excess to be used for more dowries (Anon., 1960). The administrators of the will, the Drapers' Company, though complying with the first part of the document to provide dowries to four maidens, allowed the remaining assets to accumulate over the centuries. In 1845 it was decreed that the scope of the funding be expanded to include the education of girls. Following the passage of an act of parliament in 1852, two Howell's Schools were founded in 1858, one in Llandaff, South Wales and the other in Denbigh, North Wales. Though established under the auspices of the Church of England and with an initial requirement of Anglican staff, girls of any denomination were to be accepted.

During the Taunton Commission Enquiry (see chapter 1), assistant commissioner H.M. Bompas visited schools in Wales. In the resulting 1870 report, Bompas proclaimed that the two Howell's Schools met the highest standards for girls' schools (Evans, 1990a: 68-69). Even in the 1870s there were no other academic girls' schools in the country. A particular problem was that many English independent girls' schools advertised for pupils across Wales, draining away not just potential students, but also diminishing the need for more academic schools there (ibid.: 76–7).

The Aberdare Report of 1880

In 1880 the Aberdare Committee (chaired by Lord Aberdare) was appointed by the British Government (Roderick, 2001). Its mandate was:

> ... to inquire into the present condition of intermediate and higher education in Wales and to recommend the measures which they think advisable for improving and supplementing the provision that is now, or might be made available for such education in the Principality.
>
> (Evans, 1990a: 110)

The task of the committee was to study education in general – which really meant boys' schools – few of those testifying before the commission addressed the issue of girls' schools. Though the Howell's schools were praised academically their Anglican control came under much criticism. Two presenters of note were Emily Higginson, widow of a Unitarian Minister, and Dr Frances Hoggan, née Morgan. Higginson stressed the need for teaching science to girls, while Hoggan urged consideration of a network of girls' schools modelled upon the Girls' Public Day School Company (GPDSC) (ibid.: 111–13). Hoggan subsequently authored a book, *Education for Girls in Wales* (Hoggan, 1882), which included those parts of the Aberdare Report that pertained to girls' education.

In 1881 the outcome of the hearings and deliberations was published as: *The Report of the Departmental Committee to Inquire into Intermediate and Higher Education in Wales.* Their recommendations included the founding of intermediate schools, also called County Schools, that would be co-educational and would prepare both boys and girls for public examinations.

Association for Promoting the Education of Girls in Wales

Nothing seemed to transpire from the report for several years. To keep alive the goals of girls' education in Wales, the Association for Promoting the Education of Girls in Wales (APEGW) was established in 1886 (Evans, 1990a: 132–6). Margaret Hay-Williams (Lady Verney), another of the women presenters to the Aberdare Committee, was elected the first president and honorary secretary of the organization. She was joined in her mission by Dilys Davies, formerly the first chemistry teacher at NLCS (see chapter 3). Davies, under her married name of Dilys Glynne Jones, became the vice-president of the APEGW. The first public meeting of the APEGW was held in January, 1887, with Sophie Bryant, headmistress of NLCS (see chapter 3), the speaker.

Another key figure in the organization was Elizabeth Hughes, who had taught at CLC, and from 1885 had been the principal at the Cambridge Training College (see chapter 5). In her speeches Hughes had urged the formation of a Welsh Girls' Day-School Company to parallel that of the GPDSC in England. This idea never found favour. In fact one GPDSC school had opened, Swansea High School in 1888 (Magnus, 1923: 174). The school had closed in 1895, having been unsuccessful in spreading the concept of the GPDSC into Wales (Evans, 1990a: 113).

The APEGW held meetings, organized petitions, raised funds, and distributed pamphlets (ibid.: 150–1). By 1889 the Association had 300 members. Finally, in that same year, the Welsh Intermediate and Technical Education Act of 1889 enabled some of the recommendations of the Aberdare Report to be put into practice (Evans, 1990b). With passage of the act and the rapid construction of a large number of secondary schools, both co-educational and single-sex, participation in the APEGW dwindled through the 1890s and the Association was disbanded in 1901 (Evans, 1990a: 151).

Evans has contended that, in practice, many of the wider aims of the act were ignored in favour of a strict equality of girls' education. He commented:

Any significant differentiation of the curriculum between the sexes, based on domesticity, had been regarded as incompatible with the aims of higher education for girls, which were concerned with their equal intellectual status. In Wales, as in England, there was evidence that by 1914 the curriculum of girls' intermediate and secondary schools was overwhelmingly academic. This trend had developed in spite of the advocacy of curricular differentiation by the Central Welsh Board and the Board of Education.

(ibid.: 174)

Welsh girls' science education

There was a standard syllabus for chemistry in Wales set by the Central Welsh Board and modelled upon those of the university-based English examination boards, which emphasized inorganic chemistry:

Pupils were expected to master material such as the properties of common substances, the oxidation of metals, and the experimental study of atmospheric air. Most of the questions set in the theoretical papers targeted knowledge, recall of factual information, chemical terminology, and experimental techniques, procedures and methods ... As early as 1901 examiners ... noted that the majority of schools had been provided with appropriate laboratory facilities.

(Evans *et al.*, 2008: 53)

Howell's School, Llandaff

Though we described above that there were two pioneering Howell's Schools, we could find little information on the teaching of chemistry at the Denbigh School. We do know that chemistry was taught, as in the Howell's (Denbigh) school magazine, *The Howellian*, it was reported that: 'Miss Brown, BSc Hons. Lond., comes to us from Sutton High School [GPDSC]. She teaches Physics and Chemistry' (Anon., 1925: 2). While in 1931, it was noted that, with the construction of new laboratories for biology and physics: 'The old laboratory is now used for Chemistry only' (Anon., 1930: 7). And throughout the 1920s, the magazine listed the successes of students in examinations in chemistry. With the lack of information on Howell's School, Denbigh, the remainder of the section will be devoted to the Howell's School, Llandaff.

The first appearance of science in the curriculum at Llandaff was in 1868. The Revd George Woods, father of seven girls, and an Oxford scholar, was asked to improve the academic content of the school curriculum (Sully, 2010: 40). As part of his efforts, 'he [Woods] introduced what he called "Conversational Lectures on the Physical Sciences"' (McCann, 1972: 135). The headmistress, Miss Ewing, had a strong interest in the sciences and natural history herself (ibid.: 161). During the 1870s Ewing obtained permission from the governors for the older girls to attend university extension lectures in chemistry at the University College of South Wales and Monmouthshire in Cardiff (Sully, 2010: 47). The compiler of the school's history, Janet Sully, also noted: 'In the early 1870s younger pupils, too, were introduced to Geometry, Latin and Chemistry ...' (ibid.: 40).

Chemistry teachers

The first natural science teacher was a Miss Lloyd, who was appointed in 1885. No direct information on the identity of 'Miss Lloyd' could be established. However, in that time frame there was a young woman chemist, Emily Jane Lloyd, who had links to Wales. Records indicated that Emily Lloyd had transferred from Mason College, Birmingham to University College, Aberystwyth in 1884, passing the University of London Intermediate science examination in 1887 (Rayner-Canham and Rayner-Canham, 2008: 56–7). However, it is likely she was at Aberystwyth, 1884–5, and taught science at Howell's, Llandaff, from 1885 to 1887.

It was during Miss Lloyd's time at Howell's that the first space was constructed for a chemistry laboratory, as Sully described:

> Early in 1885, Miss Kendall [the headmistress] asked for gas to be laid on in the Experimental Room for Chemistry. As there was no money for the building of a Laboratory, the former bacon loft above the stables or coach-house was fitted with gas and water and adapted for that purpose ... the only means of entry was by a ladder.
>
> (Sully, 2010: 62)

The years of climbing the ladder to the loft above the stables for science lessons came to an end in 1904 when a new chemistry laboratory was built. Thus early practical chemistry at Howell's School, Llandaff, paralleled that of many English girls' schools with a very early converted space, and then the first proper laboratory facility about the 1900s (see chapter 6).

Figure 9.1: Chemistry laboratory, Howell's School, Llandaff, c. 1890
Source: Archives, Howell's School, Llandaff

We mentioned above that we believe Miss Lloyd was the first teacher hired for natural sciences. If this was indeed Emily Jane Lloyd she returned to Mason College in 1887 to complete an external University of London BSc degree. Lloyd was succeeded at Howell's by Mary Florence Rich, who taught chemistry during 1887 and 1888 (Staff Records, Howell's School Archives).

Rich, born in Weston-super-Mare in 1865, had been educated at Haberdashers' Aske's School for Girls, Hatcham, and then commenced the chemistry degree programme at Somerville College, Oxford, in 1884. At

Oxford University, Rich was the only woman in Prof. Augustus G. Vernon Harcourt's quantitative analysis course, which in the normal manner would have required her to have a chaperone. Vernon Harcourt had been a staunch supporter of women in chemistry (Rayner-Canham and Rayner-Canham, 2009). The archivist of Somerville College, Pauline Adams, described the event:

> In her second year, Miss Rich was the only woman in a class of eight studying quantitative analysis in Mr Harcourt's laboratory at Christ Church. Mr Harcourt, whose refusal to lecture separately to women students had hastened their admission to university lectures, had equally strong objections to allowing chaperones in his laboratory; an arrangement was therefore arrived at by which, at times when Miss Rich was in the laboratory, Miss Seward [see online Appendix 2] carried out research with Mr W.H. Pendlebury on the rate of chemical change.
>
> (Adams, 1996: 40)

In Rich's third year at Somerville, life was not much easier, as Adams noted: 'When in her third year Miss Rich was at last able to devote herself exclusively to chemistry she found herself "frightfully handicapped by being the only woman; always having to depend on someone to accompany me to lectures or library …"' (ibid.: 40).

Rich completed the Oxford honours chemistry degree requirements in 1887, then became a 'steamboat lady' (see chapter 5) to obtain an MA from Dublin. The position at Howell's, Llandaff, was her first teaching appointment. She stayed only until 1888, obtaining a teaching position at Grantham Ladies' College from 1889 to 1892, and then at Wimbledon House School (later Roedean School) from 1892 to 1905.

Rich then left Roedean in 1905 to set up her own girls' school, Granville School, Leicester, of which she was principal from its opening in 1906. In Rich's obituary in the journal *Nature*, it was noted that the school 'acquired a considerable reputation …' (Fritsch, 1939: 845). Rich had to resign in 1923 due to ill health. However, she then became an honorary assistant in the Botanical Department of Queen Mary College, London. Previously, her hobby had been freshwater algae, and the academic position now enabled her to pursue her interest as serious research, which led to recognition of her contributions. Rich died in 1939 (Child, 1939: 8).

The longest-lasting chemistry teacher at Howell's School, Llandaff, was Alice Georgina Winny. Winny, born in 1865, had been educated at King Edward VI School, Birmingham, and then spent three years studying science at Newnham College. Appointed in 1889, like so many of the women science teachers, she devoted her whole life to teaching at one school. This point was emphasized in the retirement announcement from the principal, Miss Trotter:

> As we all know, Miss Winny is leaving us this term, having been Science Mistress of Howell's School for 41½ years. She came here very young from Newnham, and though, as the years went on, many schools in England and Wales would have been glad to welcome her as their Head Mistress, she was devoted to our school, to her teaching, her scientific work, and she refused to go elsewhere. She gave all her working life to this School ...
>
> (Trotter, 1930: 6)

Upon Winny's death in 1944, Trotter commented: 'Many of the girls doing work in Medicine and Science today received their first inspiration from her profound knowledge and her love of scientific truth' (Trotter, 1944: 9–10). While one of her former students had a more personal perspective on Winny: 'We were terrified of her. We adored her. We shall tell our grandchildren about her. That's the sort of woman she was – an unforgettable influence upon generations of Welsh schoolgirls. Thank you, Miss Winny. They were fortunate generations' (Morgan, 1944: 14).

Winny was assisted from 1918 until 1922 by Ivy Rosina Davis (see online Appendix 2). Upon Winny's retirement in 1930, Emily Mary Willey was hired. Willey taught chemistry from then until her retirement in 1946 (see online Appendix 2).

Student chemistry activities

Only one chemistry-related activity could be found in *Howell's School Magazine*: an outing to Cardiff to hear three lectures, the first two given by Sir W.L. Bragg and the other by his son, Prof. W.H. Bragg. The following report was given by Mary McFarlane and Irene David:

> Sir W.L. Bragg dealt with the subject [From the Atom to the Solid] in a very breezy and intensely interesting manner. He explained that this subject, on which he and his son have been researching for the past twenty years, was an entirely new branch of science. It consisted fundamentally of the examination of crystal structures by X-ray and he showed us some remarkable slides of the arrangement of atoms in molecules of various compounds. It surprised us to see how amazingly accurate the organic chemists had been in their conjecture of the structural formulae of these

compounds ... The interest of the lecture was further increased by several practical experiments.

Having enjoyed hearing the father so much, we eagerly awaited the son's (Prof. W.H. Bragg's) lecture. The subject was "The Application of Crystal X-ray Analysis to the Sciences." ... The lectures have helped us a great deal ...

(McFarlane and David, 1935: 34–5)

City of Cardiff High School for Girls
The Cardiff Intermediate School for Girls had been founded in 1895 as an examination-entry, fee-paying girls' school following the passage of the Welsh Intermediate Education Act of 1889. It was renamed the City of Cardiff High School for Girls in 1911. The purpose of the school was to offer girls' education with excellence in arts and in science.

Chemistry laboratories
The school quickly outgrew the original site and plans were made for a new building. In these plans a chemistry laboratory was included. Whether girls needed chemistry became a contentious issue as was described in the history of the school:

In those days, the Charity Commissioners [see chapter 1] were the final authority whose sanction must be received for any building. Our new plans included a chemical laboratory – the Charity Commissioners returned the plans with the laboratory struck out, giving the reason that chemistry was not a necessary subject for a Girls' School! The Governors were anxious to avoid delay and did not stay to argue the point. The room was labelled "Sewing Room," and so approved; the dimensions somewhat curtailed. At a later date (when perhaps it had occurred to the Charity Commissioners that even cookery and hygiene would be more intelligently understood with some knowledge of chemistry) sanction was given for fitting up the laboratory.

('M.C.', 1924: 18)

THE LABORATORY

Figure 9.2: Sketch of the first chemistry laboratory at the Cardiff High School for Girls

Source: Anon., 1924: 46

This laboratory remained in use until 1930, when extensions, including new science facilities, were added. It was reported by Catherine Carr 'After the [opening] meeting, parents went to view a very fine Chemical Laboratory indeed, with windows on both sides of its long length and views over the hills, and with adjacent Balance and Dark rooms' (Carr, 1955: 43).

Chemistry teachers

Florence Gibson was the first chemistry teacher at the school. Unfortunately, we have been unable to find any details about her personal background. In the history of the school, Gibson is mentioned as follows:

> Miss Gibson, who was to be identified with Chemistry more than any other subject, was appointed to teach Botany and some Mathematics; but it was not long before she was demonstrating the rudiments of Chemistry and teaching General Science in the room labelled 'Sewing Room', and agitating for the equipment that was to turn it into a small cramped laboratory, but one in which the girls could experiment themselves.
>
> (ibid.: 123)

Gibson's successor was Margaret E. John (ibid.: 124). John, born in 1898, had been admitted to Howell's School, Llandaff, on a scholarship at age

13 and she subsequently became head girl. She entered Bedford College in 1917, obtaining a BSc in 1921 (Staff Records, City of Cardiff HSG Archives). John taught at King's High School for Girls, Warwick, from 1921 until 1924. She then returned to Bedford College to undertake an MSc in inorganic chemistry with Prof. J. Spencer that was completed in 1926.

In 1926 John obtained her appointment as chemistry teacher at Cardiff High School. A former student, Alicia Lewis, reminisced about John:

> I can always see Miss John walking the length of the Chemistry Laboratory, that Palace of Experiment on the top floor of the new building, and seating herself on the dais at the end. Miss John taught us and chaffed us through Chemistry with sureness of method and deftness of touch which makes me try to reflect her lab. of those days in my kitchen of these.
>
> (Carr, 1955: 124)

John retired in 1946, after teaching 20 years at Cardiff High School for Girls. She moved to Australia and sent back to the Howell's School, Llandaff, detailed accounts of her life in the Antipodes (John, 1949).

The field club and science club

It was Gibson who founded the field club in 1901. The club had a membership limited to a range of 25 to 30. Each year new members were admitted, but also the former members had to apply for readmission ('F.G.' 1924: 27). When Gibson left in 1926, the club came to an end (Carr, 1955: 123).

It was not until 1930 that a science club was instituted. The formation of the club by Miss John was described in the *City of Cardiff High School for Girls Magazine*:

> It had been felt for a long time that some sort of Scientific Society was needed in the school, but the old laboratories were not suitable for the purpose. In the Autumn term of 1930, when the new laboratories had been firmly established, Miss John called a meeting of the Science Staff to consider the formation of a Science Club. It was decided that the Science Club should be open to all the Science Staff, all girls in forms VIA and VIB, and the science girls in the VA forms, and that two meetings should be held every Autumn term, one and an expedition in the Easter term, and one open evening in the Summer term. Miss John was elected President, and Miss Rees, Vice President.
>
> (Anon., 1931–2: 46)

Then it was noted that at the first meeting: 'Miss John gave a lecture on 'Liquid Air' which was much appreciated' (ibid.: 46).

A chemistry story

In chapter 8 we provided examples from school magazines of short stories in which elements were anthropomorphized. In a similar manner, a student writing in the *City of Cardiff High School for Girls Magazine* described the chemical reaction between hydrogen gas and (violet) iodine vapour to give hydrogen iodide. The account commenced: 'Great interest attaches to the charming wedding which was celebrated at the Llab. Oratory yesterday, when Miss Io. Dyne was married to Mr Hy. D. Rogen, the eminent reducing specialist' (Anon., 1927: 28).

Pontypool County School for Girls

The County School for Girls, Pontypool, was also established as a result of the passage of the Welsh Intermediate Education Act of 1889. Founded in 1897, like the City of Cardiff High School for Girls, the Pontypool school aimed at academic excellence. As we report below, the founding headmistress, Miss Dobell, was the driving force behind the first chemistry laboratory.

The second headmistress, Miss Jones, was equally supportive of science for girls. In *The History of the County School for Girls, Pontypool*, it is noted:

> The war years 1914–18 saw a rapid extension of such opportunities for women, especially in the realms of science. Miss Jones, keenly aware of this trend, gave great significance to the science side of the curriculum. She realised that Science teachers would be needed for the new Secondary Schools which were being built and that women scientists would be wanted in industry and scientific research.
>
> (Anon., 1947: 32)

Chemistry laboratories

In chapter 6 we noted that some girls' schools believed that they were one of the very few to have a chemistry laboratory before 1900, even though they were one of many. Pontypool was no exception. In the history of the school, it is commented:

> At first there was no opportunity for the study of Science, but in June 1898 the large room above the present dining-hall was

> equipped as a Chemistry laboratory. The fact that this came about
> at an unusually early date for girls' schools bears witness to Miss
> Dobell's determination to give to the girls the same opportunities
> as those enjoyed by their brothers.
>
> (ibid.: 19)

The early school inspector's reports have survived. For example, under the section discussing practical chemistry, the report of 1908 includes: '… more volumetric exercises should have been done, and greater accuracy is desirable; [qualitative] analytical work good; organic work rather weak in theory, and more attention should be paid to the purification of the products' (Anon., 1908: 5).

In 1936 a wing was added to the school that contained a new well-equipped chemistry laboratory (Anon., 1947: 53). Though the laboratory might have been new, the aromas were the old familiar ones. In 1938, a student, Mavis Sancho, made a plea in the school magazine, *The Dragon*: 'We fervently hope the Government will hurry up with those gas-masks, because the smells they manufacture in the Chemistry Laboratory …' (Sancho, 1938: 15). Three pages later, a fake response is given by a student under the pseudonym of 'I Smelt B.A.D.E.G.G.S': 'Dear Madam, Messrs. Gassum and Co. are pleased to recommend to you a fashionable line in new-model gasmasks. We assure you that they would immeasurably relieve the distress of the forms in the vicinity of the Chemistry Laboratory' ('I Smelt B.A.D.E.G.G.S', 1938: 18).

Chemistry teachers

As we mentioned in earlier chapters, staff registers did not commence until one or two decades after the founding of a school. The records of the County School for Girls, Pontypool, were no exception and only began in about 1905. The first chemistry teacher was listed as Maude Samson Carter (Staff Register, Pontypool County School Archives). Born in 1885, Carter was educated at Redland High School for Girls, Bristol, from 1896 until 1903, and that year she entered University College, Bristol (later the University of Bristol). She left University College, Bristol in 1907, taking up an appointment as elementary science and chemistry teacher at Pontypool in 1908. Carter left the school in 1910 to return to the University of Bristol to complete her degree. We could find no subsequent information on her.

Carter's successor was Norah Dorothy O'Flyne. O'Flyne was born in 1888 and attended a series of local schools in Cardiff before entering University College of Wales, Cardiff, in 1905. She graduated in 1909 with an honours degree in chemistry (Staff Register, Pontypool County School

Archives), being hired at Pontypool in 1910 for: 'Whole of the teaching in Chemistry'. O'Flyne resigned in 1915 to train to become a medical doctor at the Royal Free Hospital for Women (previously named the London School of Medicine for Women).

The principal chemistry teacher from 1916 until 1945 was Margaret K. Turner. Turner, born in 1880, was educated at Orme Girls' School, Newcastle-under-Lyme (Staff Register, Pontypool County School Archives). She completed an honours BSc in chemistry at University College of Wales, Aberystwyth (UCWA) in 1909. Turner held a series of short-term teaching positions after graduating. Then, at the beginning of the First World War, she was hired as a demonstrator in chemistry at the UCWA. In the history of the chemistry department of UCWA, it was reported that: 'The advent of war in 1914 caused considerable disruption ... Miss Margaret Turner and, later, J.B. Whitworth took up posts as demonstrators and under the direction of the professor carried out large-scale preparations of intermediates for certain essential drugs' (James and Davis, 1956: 571–2).

In January 1916 Turner was appointed as chemistry mistress at Pontypool (Staff Records, Pontypool County School Archives), being promoted to senior science mistress in 1934. She retired in 1945 and died in 1961. In the obituary in the school magazine, *The Dragon*, it was commented: 'She was an enthusiast in the teaching of her subject and introduced us to many interesting "sidelines," such as the crystal-growing competition in the Eisteddfod' (Wilks, 1961: 6).

Welsh branch of the Association of Women Science Teachers

In chapter 5 we described the Association of Women Science Teachers (AWST). We noted that there were a few regional branches, and the most active of these was the Welsh branch (Lewis, 1966). The branch was founded in 1921 and continued to be active until the merger of the AWST with the Science Masters' Association in 1962. The meeting at which the formation of a Welsh branch was proposed was held at the Cardiff High School for Girls, and the subsequent meetings alternated between Cardiff and Swansea. During the inter-war period, the attendance was typically between 15 and 20 members.

Among the early committee members was the chemistry teacher, Miss Gibson of City of Cardiff High School (see above), while later there were the chemistry teachers Miss John of City of Cardiff High School (see above) and Miss Turner of the County School, Pontypool (see above).

From 1933 onward the most prominent chemistry teacher in the organization was Eluned Leyshon. Leyshon taught at Glanmôr Girls' Secondary School in Swansea. Born in 1909 in Swansea, she completed an honours BSc in chemistry at University College, London (Anon., 1948). Leyshon started teaching at Glanmôr in 1931, becoming head of science in 1948. She died in 1982.

According to the *Committee Minute Books*, most of the meetings of the Welsh branch involved lectures or discussions. For example, on 12 December 1922, a presentation was given by a Miss Abbott at City of Cardiff High School on 'Teaching of Science in Girls Schools'. There were also visits to company works, such as the Mellingriffin Tinplate Works in 1927, and to university research laboratories. In addition, the Welsh branch organized scientific excursions, including a traverse of the South Wales coalfield with a geologist.

As can be inferred from the above, though it was called the Welsh branch, in fact, for meetings, it was the Cardiff-Swansea branch. This issue arose at the Annual General Meeting of 1936 when the following Motion was proposed and approved: '... that the scattered members in the schools of North Wales should be invited to meet the members of this branch at some convenient place ... Llandrindod Wells or Shrewsbury for the summer meeting' (Anon., 1936). However, there was no indication that such a meeting took place.

Though there was no coherent narrative of the teaching of chemistry at independent girls' schools in Wales, there was a similar emphasis on chemistry as an academic subject. The modelling along the lines of academic English girls' schools was apparent through the indirect involvement of individuals, such as Dilys Davies of APEGW, Sophie Bryant of NLCS, and Elizabeth Hughes of CLC. For chemistry teaching, the influence was more direct, with several of the teachers having attended chemistry-active English independent girls' schools (given in parenthesis), such as Alice Winny (King Edward VI) and Emily Willey (James Allen's Girls' School) of Howell's School, Llandaff; and Maude Carter (Redland High School) and Margaret Turner (Orme Girls' School) of Pontypool County High School for Girls.

Chapter 10

Chemistry at some Scottish girls' schools

'Dear Somebody, I suppose you are up to your necks in H_2O, and I suppose your feet will show traces of moisture. Doubtless in the circumstances, you will welcome the SOK_2S which I enclose. Also the X_3 which go along with them to speed them on. Ever yours, – '

(Paterson, 1915: 33)

Though the educational system in Scotland was quite different from that in England and Wales, the interest in chemistry was equally apparent north of the border. In the quotation above, the students incorporated chemistry content into notes to accompany the socks they had knitted for soldiers in France during the First World War.

Our account of girls' academic education in Scotland focuses on four independent girls' schools: two in Edinburgh and one each in Glasgow and St Andrews. The small number is, in part, due to the predominance of co-educational schools in Scotland. However, an additional reason was the significantly lower proportion of Scottish girls' schools that, compared with English girls' schools, were willing to allow us access to their school and/or archive.

Girls' education in Scotland

The beginning of the academic education of girls in Scotland dates back to the early nineteenth century. Much of our account of this initial period has been gleaned from the comprehensive research of Lindy Moore (Moore, 2003) and to her we express our thanks.

From about 1815 until 1835 the offering of specific subjects by individual teachers at their home was a particular feature of education for Scottish girls, as a Mrs Furlong commented in 1855:

In every street might, at the period to which I refer [1815–35], be seen some delicate girl, hurrying from class to class (according as each teacher found their partizans), which were held at their respective houses. These young creatures were generally encumbered with a load of books, and the time lost in their endeavour to make the respective master's classes hours meet, being found at length an evil which called for some better plan ...

(Furlong, 1855: 6)

Progress towards the formation of schools occurred largely, and independently, at the two metropolises of Edinburgh and Glasgow. Scotland did not have the equivalent of England's Miss Buss or Miss Beale (see chapter 3); instead the young ladies' institutions were administered by men, and men made up the majority of the teaching staff (Moore, 2003: 256–7).

In Glasgow the first girls' school formed was Queen's College in 1842: '... the most academically and socially ambitious of all the institutions ...' (ibid.: 252). But it ceased to exist four years later. The Glasgow Institution for the Education of Young Ladies (GIEYL) opened in 1845 and survived for about 40 years, while the West of Scotland Institution opened in 1848 and was still taking students into the 1890s.

In Edinburgh the first girls' school was the Edinburgh Ladies' Institution for the Southern Districts (ELISD) in 1833, followed the next year by the Scottish Institution for the Education of Young Ladies (SIEYL). According to Moore, SIEYL was opened 'by teachers influenced by the contemporary movement for educational reform, especially relating to science, pervading Scottish society' (ibid.: 251). Teaching was accomplished by 'the group of specialist masters, covering every subject normally taught to girls with the addition of science ...' (ibid.: 251). A third school, the Edinburgh Institution for the Education of Young Ladies, opened in 1835.

Chemistry teaching to girls in the early 1800s

At SIEYL it was Dr David Boswell Reid, physician and chemist, who was the enthusiastic chemistry teacher (ibid.: 261). Dating from 1835, a report (though it is more of a prospectus) on the courses offered by SIEYL includes a wealth of information on chemistry. In the chemistry section, authored by Reid, chemistry is extolled as a vital part of modern (early 1800s) life:

... *Chemistry* has been marked by the rapidity of its progress, the singular nature of the new relations it has disclosed, and by results so important, that, wherever we turn our eyes, its beneficial influence may be seen ministering to the wants, and improving the conditions of man. It has thrown a new light on the whole economy of nature, and explained a thousand mysterious operations of which formerly we were completely ignorant.

(Anon., 1835: 26–7)

The report noted the diligence of the young women: 'In the Chemical Class, the ladies answered many questions, wrote exercises, and practised, under Dr Reid's superintendence, many useful experiments' (ibid.: 8). Parents were urged to allow their young children to attend the science classes, including chemistry:

> What profit, it may be asked, can a child of ten or twelve years derive from the geological, chemical, or botanical lectures. It is not expected that these young ladies should, after hearing *one* course, be either chemists, botanists, or geologists. The purpose is completely attained, if, at the outset, their attention is awakened to the contemplation of the wonderful works of nature; and it is the fact, that these scientific lectures, accompanied as they are by many interesting experiments, have been listened to by the junior pupils with eagerness and avidity.
>
> (ibid.: 8–9)

The report provides the full content of the chemistry course which showed that, for its time, it was very comprehensive and thorough. The contents match with those of Reid's own textbook, *Rudiments of Chemistry; With Illustrations of Chemical Phenomena of Daily Life* (Reid, 1836):

> ATTRACTION – CHEMICAL ACTION – Its universal Agency in sustaining the general Economy of Nature – Sketch of the Atomic Theory – Crystallization.
>
> HEAT – Communication of Heat – Liquefaction – Vaporization – Steam – Distillation – Spontaneous Evaporation – Means of producing High and Low Temperature – ELECTRICITY – GALVANISM.
>
> Chemical History of Elementary Substances – OXYGEN – Extensive Chemical Agency – Deflagration and Detonation – Hydrogen – Water – Atmospheric Air – Ventilation – Combustion – Sulphur – Phosphorus – Carbon – Carbonic Acid Gas – Coal and Oil Gases – Fire damp – Chlorine – Iodine.
>
> ALKALIS – Potassium – Potassa – Nitre – Gunpowder – Soda – Kelp – Barilla – Rock-salt – Mode of preparing Salt – Ammonia.
>
> EARTHS – Lime – Plaster of Paris – Chalk – Baryta – Strontia – Silica – Sand – Manufacture of Glass – Alumia – Clay – Bricks – Pottery-ware – Porcelain.

COMMON METALS – Reduction of Metals from their Ores – Iron, Steel, and Ink – Lead – Copper – Zinc – Tin – Antimony – Arsenic – Bismuth – Chrome – Manganese – Gold – Silver, and Marking Ink – Mercury – Platina.

Vegetable Chemistry – Gum – Sugar – Starch, etc. – Colouring Matter – Bleaching – Dyeing – Action of Mordants – Calico printing – Oils – Soap, etc.

Animal Chemistry – Animal Substances – Albumen – Fibrine – Gelatine – Respiration – Digestion, etc.

(Anon., 1835: 29)

The SIEYL Report of 1838–9 adopted an equally enthusiastic tone for the teaching of chemistry to young girls:

In conducting the course on CHEMISTRY, the Lecturer introduced the history of the Atomic Theory, and some other departments of the science which have occasionally been regarded as too abstruse for a junior elementary class; but the answers which he received to the questions proposed during the lectures gave him full and satisfactory evidence that the zeal and intelligence of the pupils had not been over-estimated. In addition to the lectures, a great number of young ladies attended courses of Select Experiments, which were arranged according to a plan explained in Dr Reid's "Rudiments of Chemistry" ...

(Anon., 1839: 8–9)

Moore has shown that the teaching of chemistry to girls was not just undertaken at SIEYL, but promoted at several of the new academic institutions for girls in Edinburgh and Glasgow (Moore, 2003: 261–2). As an example, the Edinburgh rival of SIEYL, ELISD, was equally boastful of its science offerings, including chemistry. Moore quotes a Robert Anderson who described how, at a variety of girls schools during the 1830s, he had covered the topics of:

... a general view of the physical phenomena of the Mineral, Vegetable, and Animal Kingdoms, with an account of some of the leading facts in Chemistry and Natural Philosophy. The instructions, which are illustrated by Drawings and Experiments, are delivered in the form of familiar lectures, of which the pupils will take notes in the Class.

(ibid.: 261)

On the other side of the country, GIEYL offered physical science from 1847 and even chemistry in 1855, though by the mid-1860s science had disappeared from the basic curriculum (ibid.: 262).

The Argyll Report of 1865–8

The 1860s had been a turning point for Scottish education. The 1861 Education Act had removed the provision that teachers in Scotland had to be members of the Church of Scotland or subscribe to the Westminster Confession of Faith (that is, a list of fundamental Protestant doctrines). Then in 1864 the Argyll Commission was appointed. Just as the Taunton Report had played a major role in the future of academic girls' schools in England, so the Argyll Commission Report had a major influence on education in Scotland (Cruickshank, 1967). The Commission, chaired by George Douglas Campbell, eighth Duke of Argyll, had a mandate to investigate all types of schools and make recommendations for the future of primary and secondary education in Scotland.

In the four volumes of the report, published between 1865 and 1868, the Commissioners were particularly scathing of the paucity of secondary education in the country. They also emphasized the need for endowments for new and existing schools. The Commission specifically commented on the lack of opportunities for girls and the need for organized courses in science for girls. As a result of the Commission's Report, the 1872 Education (Scotland) Act was passed in Westminster, resulting in the massive development of secondary education facilities. Ten years later, the Educational Endowments' (Scotland) Act was approved, assisting with the funding for independent girls' schools.

St Leonards School, St Andrews

It was at the end of 1876 that the St Andrews Ladies Educational Association proposed the formation of a school for girls with the highest possible academic standards (Grant, 1927: 1–13). At its opening it was called St Andrews School for Girls, and not until 1882 was it renamed St Leonards School. As first headmistress the association chose Louisa Innes Lumsden, who, in 1872, had graduated from Benslow House, Hitchin, the precursor of Girton College. Lumsden had been one of the first women students to pass the Cambridge Tripos Examination. After completing her studies, she had obtained a teaching position at Cheltenham Ladies' College (CLC).

Also teaching at CLC was Jane Frances Dove. Dove had been educated at Queen's College, Harley Street (see chapter 3) and, like Lumsden, at Benslow House. One of Lumsden's first actions as headmistress at St Leonards was to offer the position of science teacher to Dove. With the leadership of Lumsden (and later, Dove), St Leonards saw itself less a part of the Scottish education system and more a Scottish outpost of the academic (CLC) English girls' boarding school model.

Chemistry teachers and laboratories

From its founding in 1877 chemistry was taught at the school. Julia Mary Grant, the author of the St Leonards school history, commented:

> Science from the beginning formed a regular part of the school curriculum. The sixth form were taught chemistry by Miss Dove, who was one of the first women to take the Natural Science Tripos at Cambridge, and the work soon expanded when, after the removal to St Leonards [House in 1882], a laboratory was arranged.
>
> (ibid.: 43)

A source of pressure for teaching chemistry was the opening of London University to women in 1878 (Harte, 1979), for at the time high-school chemistry was a necessary prerequisite for admission to science degrees. This point was made explicitly by Miss Dove in a letter to the administrative council of the school in 1880:

> A letter was read from Miss Dove regarding the necessity of procuring certain chemicals and chemical apparatus with a view to the preparation of pupils for the Examination of the London University. It was agreed that a sum not to exceed £15 might be expended for this purpose.
>
> (Anon., 1880: n.pag.)

It is clear from subsequent minutes that Miss Dove intended to purchase whatever was necessary, ignoring the limitation of the council:

> The account for the chemical apparatus, etc. obtained by Miss Dove amounting to £22.15.10 was laid upon the table. It was explained that every economy had been used in the purchase of the necessary materials and in the circumstances it was agreed that the full amount should be paid by the Council, although exceeding the sum formerly agreed upon.
>
> (Anon., 1881: n.pag.)

Following Miss Lumsden's retirement, Miss Dove was appointed headmistress in 1882, while continuing to teach chemistry. Dove was very concerned that the standard of chemistry education be of the utmost rigour and comparable to that of the best boys' schools. To this end she had Thomas Purdie, professor of chemistry at St Andrews University, as an external examiner. In 1885, his report included the following comments:

> The answers given show that a great proportion of the girls have acquired for their age a large amount of actual knowledge of facts, and, what I take to be still more important, habits of accurate observation, and an exact and clear way of thinking.
>
> Forms Lower VI and V: ... the chemistry of the non-metallic elements.
>
> Forms IV and Lower IV: ... the chemistry of some of the more common elements and their compounds.
>
> Forms III and Lower III: ... the chemistry of common things.
>
> (Purdie, 1885: n.pag.)

Dove left St Leonards in 1895 to found Wycombe Abbey School and, as we mentioned in chapter 6, she ensured that chemistry was a key part of the curriculum there, too.

Figure 10.1: The old chemistry laboratory at St Leonards, c. 1908
Source: Archives, St Leonards School

The next chemistry teacher of note was H.P. Auld, who arrived in 1902. Auld had received an Honours BSc from Victoria University, Manchester (later the University of Manchester). However, it was not until 1908, with the appointment of Mary Bentinck Smith as headmistress, that chemistry was allowed to flourish again. The author of a later history of the school, J.S.A. Macaulay, commented:

> The new Head Mistress [Bentinck Smith] was well aware of the changing situation brought about by the increasing demand for the economic independence of women as the struggle for the suffrage entered its last stage. The number of girls demanding careers and anxious to proceed to university was increasing, and Miss Bentinck Smith was determined that it should increase further still, and that St Leonards must equip them to qualify not only by providing a liberal education but also by enabling them to pass the necessary examinations.
>
> (Macaulay, 1977: 15)

The first step Bentinck Smith took was to request a new science wing including a well-equipped chemistry laboratory. In the council minutes of 1908 it is reported: 'New Science Building – Miss Auld had visited laboratories in a number of schools in England, and she and the chairman were in communication with the Architect, who was preparing preliminary sketch plans' (Anon., 1908: 99). In a speech accompanying the opening of the new science facilities, Bentinck Smith thanked those who had made the facilities possible:

> ... especially Professor Purdie, whose expert advice had been invaluable, Dr Murray, who had generously devoted much time and attention to the plans and equipment of the laboratories, and Miss Auld, who had worked literally night and day in the same cause.
>
> (Anon., 1910a: 28)

In the school magazine, the *St Leonards School Gazette*, students added their own praise:

> Only those who have worked under both old and new conditions can know the joy of working in such a delightful building ... The building comprises two large laboratories each 34 feet by 30 feet, a lecture theatre to seat 60 girls, a room for the preparation of experiments by mistresses ...
>
> (Anon., 1910b: 33–4)

Figure 10.2: Chemistry students undertaking a steam distillation in the chemistry laboratory, 1920

Source: Archives, St Leonards School

Auld left St Leonards in 1918 (Anon., 1918: 40) to become headmistress of the College for Daughters of Ministers of the Church of Scotland and of Professors in the Scottish Universities in Edinburgh (subsequently renamed the Esdaile School). Her successor as chemistry teacher was Ethel Benn who taught chemistry at the school until her retirement in 1946 (see online Appendix 2).

The science club

The St Leonards School science club was formed in 1901. A report appeared in the school magazine:

> Last Term a Science Club was formed in School for the purposes of fostering the growing interest in Science and furthering research. The club started under auspicious circumstances with … fifty members and £3 10s from a former club, now obsolete. The Society hopes to hold at least three meetings a Term, which promise to be of a very interesting and varied character. All in whom there glows any spark of scientific interest are invited to pay the annual subscription 1s and become members of the Society.
>
> (Anon., 1901: 483–4)

One wonders whether the mention of a 'former club' might have been a field club, in view of the science club 'inheriting' the funds. There is a reference by Macaulay to a gas-works episode in the mid-1880s, after the move to the St Leonards buildings:

> ... all these helped to make the girls feel that they were adventurers in a new and unknown land where everything was possible. Occasionally this spirit of adventure took mistaken forms – ... climbing ... on to the top of the receiver of the gas-works – but the satisfaction gained from such episodes was not sufficient to make them popular ...
>
> (Macaulay, 1977: 7)

Park School, Glasgow

The Park School, Glasgow, owed its existence to the success of the Girls' Public Day School Company (GPDSC), and a desire to see a school founded on the same principles in the city. In 1879, a public meeting was held in Glasgow to consider whether there was support for the formation of a similar company to establish a school, or schools, according to the GPDSC principles in the area. With strong local support, The Girls' School Company Limited, Glasgow was formed. The Park School already existed as a private boy's school, and it was purchased and remodelled as a girls' school, opening its doors in 1880 ('M.M.', 1930: 9–11).

The first headmistress was Georgina Kinnear. Kinnear had been educated at the CLC under Beale. She had also attended some of the Schools Inquiry (Taunton) Commission hearings in London (see chapter 1) and heard Miss Buss speak of the need for academic girls' schools. Louisa Lumsden, headmistress at St Leonards, had met Kinnear at the CLC and was very impressed by her. Lumsden had persuaded Kinnear to join the St Leonards staff in 1877. Thus Kinnear was already in Scotland, and with an established reputation, when the position of headmistress at Park School, Glasgow was advertised ('A.Y.', 1930: 23–5).

Chemistry laboratory

It was only four years after the Park School opened that the addition of chemistry to the syllabus was considered an urgent issue:

> ... discussions with Miss Kinnear upon an important matter which Professor Lindsay and Professor Young raised at the Board Meeting in December 1884. They both spoke on the advisability of introducing the study of Chemistry into the curriculum

and submitted an estimate of the amount needed to fit out a Laboratory – £50 to £60. Dr Young was asked to proceed in the matter.

(Lightwood, 1981: 20)

St Paul's Girls' School and Milton Mount College (and many others) in England proclaimed that they were pioneers of practical chemistry for girls (see chapter 6) as did County School for Girls, Pontypool, in Wales (see chapter 9). Park School made a similar claim. In the 1930 history of the school, the author noted that the 1885 chemistry laboratory was 'surely one of the first laboratories to be introduced into a girls' school' ('M.M.', 1930: 11).

The second headmistress, Miss Young, recalled that the chemistry experiments sometimes had unintended consequences:

> When the enthusiastic chemists were absorbed in their investigations they were apt to leave the taps running, with disastrous results to the class-room below. Such episodes added greatly to the gaiety of the girls, but the authorities were more serious-minded, and after a very short time built a laboratory with more conventional equipment at what had been the backdoor. After various transformations this is now equipped for the advanced classes, while the juniors are accommodated in a new laboratory on the other side of the passage.

(Young, 1930: 42)

One of the former students remembered that the odour from the laboratory spread into the rest of the school: '... certain physical sensations bring old days at once back. If I could ... smell the H_2S in the passage to the hall, these memories would be more complete' ('E.C.', 1930: 58).

The only other note of the chemistry laboratory was in the school magazine, *The Park School Chronicle*, in 1915 (unfortunately none of the magazine issues between 1915 and 1927 has survived). The students noted: 'The Laboratory was painted during the summer holidays, and those who work there have no longer to contemplate a freeze of fingermarks. We are now quite sure that no school in Glasgow possesses a better or airier laboratory' ('L.S.T. and N.A.V.', 1915: 167).

Chemistry teachers

Miss Broadhurst seems to have been the original teacher of chemistry at the Park School. She was technically the physical geography mistress, but it seems that, like the early days of the CLC, chemistry was disguised as physical geography. There were several science teachers in the early years of the Park School and it is difficult to identify which ones taught chemistry. However, in 1895 Miss Beard was the first actually appointed as a chemistry teacher. She stayed for many years, though no departure date could be identified.

We know of one chemistry teacher specifically, Isabel Comber Fortey (see online Appendix 2), who taught at Park School only from 1888 to 1889: 'Which of us who knew her can forget Miss Fortey, also of unflattering tongue, her small figure moving about the laboratory trying to instil the elements of Chemistry into us?' ('M.B.W.', 1930: 48).

From 1906 until 1919, Miss Burdett was listed as science mistress. Burdett died in 1957. It was reported in her obituary:

> While in her mid-20s she was appointed to the school to reorganise the Science Department and bring it up to Leaving Certificate Standard. From the first, she made a deep impression on her pupils. Her discipline was kindly but permitted no nonsense and her teaching was such that even the least scientific acquired some knowledge.
>
> (Lightwood, 1981)

Park School Science Association

The students were particularly proud of the Park School Science Association, founded in the 1880s. One former student reminisced: 'The Park School Science Association was initiated with large ideals, meant to lead us gently into the paths of scientific knowledge and investigation on our own initiative, the shining example of Miss Somerville being held before our bewildered eyes' ('M.B.W.', 1930: 50). Mary Somerville was the renowned nineteenth-century Scottish woman scientist and science writer (Neeley: 2001).

Another student remembered that the association had been formed by Miss Broadhurst:

> ... [It] was started by Miss Broadhurst, the Science mistress. I can say very little about her, as I only remember her very vaguely, but the girls whom she taught were tremendously keen and very devoted to her. She must have had the power of inspiring them with her youthful enthusiasm and her high ideals, because, as a child not then at school, listening to my sister's and her friend's talk, Miss Broadhurst seemed to me a wonderful person, half-saint, half-fairy.
>
> ('E.M.M.', 1930: 76)

Miss Broadhust continued to play a leading role in the activities: 'Under her [Miss Broadhurst's] guidance papers were read and demonstrations given on such problems as "Why Does Ice Float?" "How is Coal Gas Made?"' (ibid.: 76–7). Sadly, as that cohort graduated, the meetings lost their science focus. By 1919, the name was changed to Park School Seniors' Association and the link with science went forever (Lightwood, 1981: 74).

Edinburgh Ladies' College

Edinburgh Ladies' College, the oldest of the Edinburgh girls' schools (Roberts, 2010), had been founded in 1694. Its existence resulted from a bequest by the businesswoman Mary Erskine (1629–1707) of 10,000 Scots merks for the 'maintenance of burgess children of the female sex' (Morgan, 1929: 100). Originally instituted as the Merchant Maiden Hospital Boarding School for daughters of Edinburgh merchant families, the school was refounded by act of parliament in 1870 as a day school under the name of the Edinburgh Educational Institution for Girls.

The school was renamed the Edinburgh Ladies' College (ELC) in 1889 (the establishment was renamed yet again in 1944 as the Mary Erskine School, after the founding benefactor). Demand for places at the ELC was overwhelming, as described by Alice Stronach in *The Girl's Realm*:

> ... to the old building some twelve hundred girls came flocking. Doctors and ministers and lawyers, farmers and shopkeepers, and even prosperous artisans (for from the first the school was democratic) sent their girls to share the advantages of Dame Margery Erskine's and the Merchant Company's joint charity. Twelve hundred girls under one roof!
>
> (Stronach, 1900–1: 479)

According to the norm in Scotland at the time, it was a headmaster who ran the girls' school with the assistance of a lady superintendent. The first headmaster was Dr David Pryde. Pryde hired the best masters to teach the senior students and revelled in the academic successes at universities of ELC students. In his autobiography, *Pleasant Memories of a Busy Life*, Pryde outlined his philosophy:

> Another aim was to give the biggest [older] girls almost the same education as that which is given to boys. In addition to what are called the female accomplishments, and English, French, and German, they were thoroughly grounded in arithmetic, mathematics, and Latin.
>
> (Pryde, 1893: 143)

Chemistry laboratory

Unfortunately no mention of chemistry laboratories could be found in any of the surviving issues of the school magazine, *Merchant Maiden Magazine*. In the school archives there is a photograph showing students in the chemistry laboratory, apparently from the 1910s (see below). The authorized history of the school by Lydia Skinner, gives only one brief mention of the construction of a laboratory about 1900: 'The Scottish Education Department's requirements from 1900 for Science and Art provision led to these funds being used in the pre-[First]-War years to provide laboratory ... facilities ...' (Skinner, 1994: 78).

Figure 10.3: Edinburgh Ladies' College chemistry laboratory, c. 1910
Source: Archives, Mary Erskine School

One of Pryde's successors, Mary G. Clarke, who was head from 1914 until 1924, was equally committed to a strong science education for girls at ELC. Clarke's determination was fuelled by her own very negative experiences of science teaching as a student at Aberdeen's High School for Girls:

> Science was only a name [at Aberdeen's High School for Girls]. The class repaired once a week for one period to a demonstration room where we watched the master set up experiments in physics which invariably failed. It would have made no difference to the pupils if they had succeeded, as no coherent explanation of them was ever given.
>
> (ibid.: 79–80)

Chemistry teachers

The first chemistry teacher was male, a Mr Stewart, assisted by a Miss Garrett:

> The Science Department was first organized in 1900 by Mr Stewart whom many girls will remember for his cheerful optimism and kindness. Along with Mr Stewart we had Miss Garrett a distinguished former pupil of Queen Street [ELC was commonly called Queen Street School], who returned to us after having held important posts in England and in Aberdeen ...
>
> (Anon., 1920: 63)

Subsequently a Mr MacLean and Mary Dunbar were hired. Both MacLean and Stewart left the school in 1914 for military service. Dunbar took over the science department in that year until her own departure in 1919 upon the return of the male teachers (Anon., 1919: 28). An eloquent parting tribute was paid to Dunbar by one of her students:

> We had many an opportunity of enjoying her fund of dry humour and her caustic wit which enlivened our dull moments and never left a sting. Everyone alike was impressed by Miss Dunbar's absolutely faithful and conscientious discharge of her duties and her unfailing accuracy and business-like procedure played no mean part in her development of scientific thought in the class-room.
>
> (ibid.)

The field club

The EDC field club was founded in 1918 and periodic reports were given in the school magazine of the lectures and excursions. In November 1923, a lecture on 'Alcohol' was given by Mr MacLean. The field club secretary, Rosemary Crerar, reported:

> The lecture was greatly appreciated by all, the experiments adding to the enjoyment – especially when someone was asked to breathe a compound of Amyl Alcohol in order to test its power of making people blush.
>
> (Crerar, 1923: 8)

Like many of the science clubs at English girls' schools the ELC field club activities included a visit to the gas works. However, rather than just straight prose, the author of this report adapted it into the style of the diary of John

Evelyn (Evelyn, 1882). Extracts of the account are given below so that the reader can appreciate the skill of the student:

> To Granton, and from thence to the Gasworks. There, on our demand, did we receive a full account and description of the manufacture of our city's gas, which is, in troth, a great convenience to us all. Coal was reduced to coke by heat ... Next to rid the gas of hydrogen sulphide it do pass over layers of iron oxide, the which, being bought by the gas company and being used by them, they do then sell at a price higher than paid by them, for the substance do then contain iron sulphide, the which is used in the manufacture of sulphuric acid. Thus to the gas company a goodly profit off iron oxide, they having both the use and the sale thereof ... A goodly scheme. Anon to the laboratory, where testing of specimens of the gas be performed. Next to the manufacture of these specimens, which scientific proceedings would have mightily interested my good friend Mr Evelyn ... A timely train to the West End, for trains do run from the factory for the workmen, and so home to lunch. And herein did more to-day than was hoped for.
>
> (Ross, 1925: 78)

St George's School, Edinburgh

As the school historian Nigel Shepley noted, the founding and early years of St George's School was greatly influenced by the pioneers in England: '... but the most significant English influence on the St George's founders was the Girls' Public Day School Company ...' (Shepley, 2008: 4).

The first headmistress of the school was Mary R. Walker (ibid.: 10–12). Walker had attended the classes of the Edinburgh Association for the University Education of Women which was the only option for higher education available to Edinburgh women at the time. She had then travelled to London where she attended the Maria Grey Training College (see chapter 5), obtaining a Cambridge Teachers' Certificate. After teaching at Maria Grey for two years, Walker returned to Scotland where she was subsequently offered a position as first headmistress of St George's School for Girls.

Chemistry laboratories

In 1938 Mary Walker reminisced about her time as headmistress, including the installation of the first laboratory facilities:

The next important step was the appointment of a specially qualified "Science Mistress." We had from the first included "Nature lessons" and Botany in our curriculum, but the Scottish Education Department had now made Science a necessary subject for the Intermediate Certificate and special provision for the candidate had to be made ... Where was there space for a properly equipped laboratory? We had to be content with a compromise. The room originally intended for exclusive use as a lecture room for the College students had long been partially utilised as a class room for the School. In 1907–8 it was, by the introduction of science benches, water-taps, and gas-burners, made available also as a provisional "Science room," and Miss Anderson (Mrs Mackworth), though not a specialist in Science, kindly undertook the teaching necessary for the certificate.

(Walker, 1938: 6)

In the St George's School archives there is an inspector's report of 1916 by the Scottish Education Department that contained a mention of chemistry:

U.5th. In Science, Botany is the favourite subject, and most attention is paid to it; but useful courses in physics and chemistry are also given. The formal science course extends over three years, and as there is a fine laboratory, pupils may attain to a high degree of proficiency.

(Anon., 1916: n.pag.)

A subsequent undated inspector's report in the St George's archives, shortly after the move to the new site, states: 'There is also a good and well-arranged science laboratory, suitable for chemistry and physics ... I also examined the Physics and Chemistry classes taken by Miss Wheeler – and got good answering from the pupils' (Anon., n.d.: n.pag.).

Just as many English girls' schools constructed new chemistry laboratories in the inter-war period (see chapter 6), so did St George's. It was the headmistress of the time, Margaret C. Aitken, previously headmistress of the GPDSC Sheffield High School, who provided the impetus. In 1928 Aitken persuaded the authorities that the school needed a new wing including a new science laboratory (Shepley, 2008: 87). Mention of the construction was also reported by a student in the school magazine, *St George's Chronicle*: 'The new laboratory is still unfinished ... Judging by surreptitious peeps into the new laboratory it is going to be equally successful. There we shall peruse Atomic Weights and Laws of Electrolysis with renewed interest' (Anon., 1930: 14).

Chemistry teachers

In 1907 the original chemistry teacher, Miss Anderson, resigned in view of her impending marriage, and Jean G. Thomson was hired as the first qualified science mistress (Anon., 1938: 6). Thomson left in 1928 to take up a position at Falkirk High School (Anon., 1929: 14). In an obituary for Thomson it was commented:

> How much the Science Department of St George's School owes to Miss Thomson can be known only to those who worked with her. But that her influence is still felt there can be no doubt. She inspired colleagues, students and pupils with her ideals of hard work, of careful preparation and devotion to scientific method. The Department increased in size during her period of service, and the addition of another laboratory made it possible to enlarge the scope of Scientific work.
>
> ('E.J.B.', 1954–5: 37)

Thomson's successor was Elinor M. Younie. Younie taught chemistry from 1928 until her marriage in 1934 (see online Appendix 2).

A chemistry poem

At St George's School a student used poetic form to describe an incident in the chemistry laboratory. The sad tale involved an experiment of the production of chlorine by the reaction between sodium chloride, concentrated sulfuric acid, and manganese (IV) oxide to give large volumes of toxic green chlorine gas.

A Disaster

A troop of damsels hurried to the lab.

To try to make some chlorinz (Cl_2).

The fair one with the golden locks was there,

Clad in a robe of purple casement-cloth,

Her lissom limbs chased in long black gym-hose,

A new St George's tie of flaming red;

Cheeks like wild rose, and eyes of speedwell blue.

A wondrous apparatus was set up –

Flask, bottles, tubes, and H_2SO_4.

"The manganese dioxide, where is it?

Run! Fetch the jar of black MnO_2!

The fair one, like a streak of lightning, swooped,
To snatch the jar with dainty snow-white hand.
Another nymph swooped too, a buxom lass,
With auburn tresses glinting thick and bright.
Both seized the jar of fateful manganese,
And either nymph was loth to give it up.
A frightful crash resounded thro' the Lab –
(Cursed be the hand that pulled the stopper out!)
But, Oh! our fair one with the golden locks,
Was like a sweep before he's had a bath.
The wretched sooty-black MnO_2
Had made the wild-rose cheeks grimy black!
The speedwell eyes were piteous with dismay –
Alas! the lovely tie of flaming red
Was all besmirched, and grievous to behold.
The class proceeded with the fair one black.
The manganese dioxide did its work,
Till ghastly fumes of chlorine filled the lab.;
Then paled each check, and rolled each yellowing eye.
The fair one sat with wan and pensive mien,
With drooping lash, and golden hair all grey,
Received with silence every scornful gibe.
I wept to see this too pathetic sight!
After the class she vanished, in distress.
To cleanse her dainty face and change her dress.

('Scientist', 1927: 14)

Despite the Scottish educational system's independence from the English, what is very apparent is that the Scottish girls' schools founded in the latter part of the nineteenth century consciously emulated the GPDSC school model in the case of day schools and CLC for boarding schools. In addition, many of the teachers and headmistresses identified in this chapter had experiences of the respective school models. Like their English counterparts, academic chemistry was the focus, so that the Scottish independent girls' schools should be seen as at least equal to that of the boys and that successful science-focused girls would be eligible for university entry in the sciences.

Chapter 11

What will the chemistry students do?

Professional life is the best instrument we possess for perfecting the individual [girl], and the laws of nature would not be interrupted because girls were not sitting waiting for husbands to find them.

(Dove, 1907: 874)

So said Frances Dove, headmistress of St Leonards School (1882–6) and of Wycombe Abbey School (1896–1910). Thus far we have looked at the fascination for chemistry of many students during their school years. But what became of them? Even in the 1880s there was concern as to the futures for academically educated young women. This worry was particularly evident for girls who had become enamoured of chemistry.

In this chapter we first survey some of the discourse about career options for these girls. Then we look at a few specific pathways chosen by young women graduates. In online Appendix 3 we provide some matching case histories. By necessity, we can only select a very few of these women, for, as we have shown elsewhere, during our time-period, more than 700 women – mainly from independent girls' schools – established themselves as professional chemists, in addition to those who chose other chemistry-related careers such as medicine, pharmacy, and science teaching (Rayner-Canham and Rayner-Canham, 2008).

Career options for young women

This book covers a 60-year time-span and we sometimes forget that this encompasses nearly three generations. The early time frame saw the real pioneers and it is those young women, and their expectations, that will largely be highlighted in this chapter.

In 1876 Clara Jowett wrote an article in the *Milton Mount Magazine* entitled 'Female education: Its aims and ends'. She waxed lyrical about the new era that had arrived: 'In these days of enlightenment, when one would think that inventions, manufactures, and literature were at a climax, one of the most striking features is the improvement and advancement of *Female*

Education' (Jowett, 1876: 66). In her list of educational objectives, Jowett included: 'Another, and very important *aim* and *end*, is to provide suitable occupation for females of the middle classes' (ibid.: 68). However, the only option that she noted was school teaching.

There were very few opportunities about that time. In an article by Alice King on 'What our girls may do' in *The Girl's Own Paper* of 1880, she offered the choices of mentoring young female servants, Sunday-school teaching, needlework, visiting the sick and elderly, and cooking. The article concluded:

> ... we would say to our girls one earnest warning word about what they may *not* do. They may *not* do any of those things which make them imitators of men; they may *not* try to break down the God-appointed fence which divides their departments ... from the department of men; by so doing they only lose their own queenliness ...
>
> (King, 1880: 463)

The need for employment opportunities was the focus of an article in an 1882 issue of the *Milton Mount Magazine*. The author offered a bleak picture of the time. In her view, the fierce competition for teaching positions meant that only the most academically gifted stood any opportunity of seizing one. She mentioned that there were some openings in the medical profession. However, she wished there were British technical institutions for the graduating Miltonians to resemble those on the Continent, such as the *écoles professionnelles* in Paris or the Lette-Verein in Berlin. At such institutions the young women were learning: '... wood, copper, and steel engraving, painting on china and glass, designing, law-copying, artificial flower-making, modelling, book-keeping, and dress-making' (Chew, 1882: 34).

New opportunities of the mid-1880s

By the mid-1880s there seemed to be more career opportunities for girls who were completing an academic school education. A book was published in 1884 entitled *What to do with Our Girls* (Vanderbilt, 1884), whose author, Arthur Talbot Vanderbilt, described the career options available for graduating girls. This book was hailed in the journal *The Literary World* as 'a most compendious work on all the possible modern employments of women ...' (Anon., 1885: 111). A wide range of employment opportunities were included, of which three were of relevance to young women with

197

a chemistry background and interest: science teaching, medicine, and pharmacy.

For a young woman wishing to follow a career in science teaching, Vanderbilt proposed she attend one of the summer courses for teachers offered by the Science and Art Department, South Kensington (see chapter 1):

> Short courses of instruction are given annually, about July, in different branches of Science, for the benefit of Teachers of Science Schools in the country. They last three weeks. About 200 teachers are admitted to them, and they receive second-class railway fare to and from London, and a bonus towards their incidental expenses of £2 each.

> A limited number of Teachers, and of Students in Science Classes, who intend to become Science Teachers, are admitted free to the Sessional Courses of Instruction in the Normal School of Science, and Royal School of Mines. They receive second-class railway fare and maintenance allowances of twenty-one shillings per week while in London ...

> (Vanderbilt, 1884: 58)

Vanderbilt recommended medicine as a career for young women with a science background, advising them to study at the London School of Medicine for Women (LSMW) which had been founded in 1874 (Thorne, 1905). He first suggested that 'Medical Women' had a role in Asia: 'In the East, where a male doctor is seldom, if ever, allowed to enter the Harems or Zenanas, they have an immense field for usefulness, which even the most prejudiced will scarcely deny' (Vanderbilt, 1884: 85). He also foresaw very specific roles for 'Medical Women' in Britain:

> There are many manufactures of a dangerous and unhealthy nature in which women are employed, such as the white lead works, &c., and, without doubt, a lady doctor would find no difficulty in obtaining an appointment in connection with them ... There is also a fine field for Medical Women in the mining and mountainous districts of Wales and Scotland. The colliers' wives, and peasantry generally, do not, as a rule, care for doctors, and even in most serious illnesses prefer to do without them.

> (ibid.: 85)

As 'Lady Chemists and Druggists', Vanderbilt believed that the career of pharmacy was particularly suited to 'daughters of country medical men, who in many instances have acquired some practical knowledge of dispensing, and of the properties of the various drugs' (ibid.: 88). He spelled out in detail the daunting quantity of chemical knowledge that a young woman planning a career in pharmacy needed to acquire:

> To recognize the ordinary chemicals used in Medicine. To possess a *practical* knowledge of the processes by which they are produced, the composition of such as are compound, and explain the decompositions that occur in their production and admixture, by equations or diagrams. To determine, practically, by means of tests, the presence in solution of the chemicals in common use, and explain the reactions which occur in each case.
>
> (ibid.: 90)

Effect of the First World War

The First World War opened up new opportunities for women, some short-term for the duration of the war, and some permanent. The editor of *The Girl's Realm* pointed out in 1915 that planning for a career was a necessity for the teen girl reader. The possibility of marriage was an unlikely prospect, considering the enormous number of male deaths on the battlefields, particularly among the middle-class officers and men: 'I address these remarks principally to girls who are facing life at the opening of their careers. The war has made us all face reality, and, not least of all, the [young] women of the world' ('Editor, The', 1915: 45). The article ended by promising that *The Girl's Realm* would continue to provide information on careers for girls.

The same year the feminist activist Elizabeth Sanderson Haldane read a paper to the educational science section of the British Association, published in *The School World*, with the same concerns, though in more detail (Haldane, 1915). She pointed out that the war had opened up whole new avenues of employment to young women. Like the editor of *The Girl's Realm*, Haldane emphasized that marriage was unlikely to be an option for most girls:

> We know too well that there must be a shortage of men with casualties reaching very many thousands. Marriage will cease to afford the normal career for very many women, and far more women will have to obtain economic independence. Parents will have to think much more seriously of the question: "what to do with our daughters." Even now the ordinary middle-class

father is anxiously applying himself to the problem, and in quite a different spirit from that of earlier days.

(ibid.: 401)

Careers in industrial chemistry

For the young women who had focused on chemistry, industrial analytical chemical laboratories provided a ready source of employment. The possibility of a career in industrial chemistry had been encouraged by the women's suffrage movement: in 1916, the following advertisement was placed in the *Journal of Education*:

> Two scholarships of £75 each are offered by *The Common Cause* (the organ of the National Union of Women's Suffrage Societies) to women who wish to qualify for positions as industrial chemists. Preference will be given to students willing to study at the Imperial College of Science and Technology, South Kensington, or the School of Technology, Manchester.
>
> (Anon., 1916a: 713)

There was an especial demand during the First World War, as was illustrated by this note in a 1916 issue of *Our Magazine: The magazine of the North London Collegiate School for Girls* about two of the former North London Collegiate School (NLCS) students: 'Gwendolen A. James is working in an analytical laboratory for a year before going on with her Science Degree work. Margarethe Mautner is analysing drugs in a Chemical Laboratory before taking up her medical studies' (Anon., 1916b: 93).

Set up in 1915, the chemical factory of Chance and Hunt in Oldbury had a ladies' laboratory, staffed and run solely by women. In the *Edgbaston High School Magazine* H.F. Fry described the facility:

> A works laboratory which is run entirely by women may, perhaps, be considered one of the most advanced products of present-day civilisation ... The staff consists of eight girls of ages ranging from fifteen to eighteen known technically as "testers," and two University women with science degrees to organise the work and train and superintend the girls. The latter were, with one exception, entirely without chemical knowledge in the beginning ... However, they were most anxious to learn and, though the task of training them must have been arduous at first, they rapidly gained a working knowledge of beakers, burettes, Bunsens, and so forth ... one did the acidity in the caustic liquors and other

tests in connection with the alkali plant, another estimated the moisture and ash in the different kinds of fuel used on the works; the various ores from the copper process kept three girls busy, the zinc plant two, and so on.

(Fry, 1918: 33)

In 1916 the Edinburgh Ladies' College magazine, *The Merchant Maiden*, had an article on opportunities for women, and it highlighted: 'The study of chemistry, especially in its application to industrial pursuits, is one of ever-increasing importance' (Clarke, 1916: 4). The article continued with a quote from the principal of Heriot-Watt College: 'Chemistry is also a subject for which girls are peculiarly fitted ... Before entering upon such a course, a girl should obtain her Leaving Certificate and then should enter as a student for the Examinations of the Institute of Chemistry' (ibid.: 5).

In the same issue of *The Merchant Maiden*, the reports on past students included two who had taken up positions in analytical chemistry:

Maisie B. Robertson has for the present forsaken her scientific studies at the University here and is engaged meantime in an explosives factory in the South of England in chemical analytical work, the nature of which, the Censor says, may not be divulged.

Helen Lumsdaine writes: "I have been working since September as an analytical chemist in the Cement Works at Cousland. This may not sound very attractive, but I have found the work quite interesting. The employment of lady chemists is an experiment which one or two firms are trying on account of the war – another occupation thrown open to women."

(Anon., 1916c: 25–6)

Alice Raisin, a former student at NLCS, was convinced of the potential employment for girl graduates in analytical chemistry. She wrote back to the headmistress, Dr Bryant, of her own experiences as an assistant chemist with a chemical manufacturing company. Part of the account was subsequently published as part of an old girls column in *Our Magazine: The magazine of the North London Collegiate School for Girls*:

They have not had a woman in the laboratory before and were obviously rather nervous over the experiment, but they seem quite satisfied with me, and have just asked if I can find them some one as a junior assistant to take the place of a boy who has to leave to join the Army.

> I consider it would be a good opening for any one keen on chemistry, who was unable to go to College and would otherwise drift into commercial work through lack of other opportunities ... I feel sure there is an immense opportunity for women in this kind of work. Just now there are far more vacancies than women to fill them, and after the war prospects are likely to be far more promising even, as one hears on all sides of extensions and improvements that different firms are going to make then.
>
> (Anon., 1918: 13)

Analytical chemistry continued to prove an avenue of employment for young women over the subsequent decades. To illustrate the potential, in 1920 Emily L.B. Forster, a lecturer in chemistry at the Westminster College of Pharmacy, authored a specific book for women chemistry students: *Analytical Chemistry as a Profession for Women* (Forster, 1920).

Alicia Maunsell, a graduate of Cheltenham Ladies' College (CLC) reported back in 1938 to *The Cheltenham Ladies' College Magazine* of her experiences at the food processing facilities of J. Lyons and Co. Ltd. When hired, Maunsell was told that 'women chemists were paid less than the men because they could not be of equal use to the firm, being unsuitable for supervising tests in the factory or for holding positions of authority' (Maunsell, 1938: 5). Despite her own success, having been appointed to supervisor rank after ten years with Lyons, Maunsell noted that there were still few opportunities for women:

> Those in authority realise the greater likelihood of women leaving after a few years' service; also many of the men chemists in senior positions are apt to give opportunities (when they arise) to the junior men chemists rather than to the junior women, who thus may lack chances to show their capabilities. This attitude persists partly from prejudice and partly because some women do not take their jobs seriously.
>
> (ibid.: 5)

Nevertheless, Maunsell recommended industrial analytical chemistry as a career for girls, describing why she found it so rewarding:

> ... if you are glad to turn your scientific knowledge to practical account, if you can derive pleasure from routine work done quickly and accurately, or if, for instance, when the canning of strawberries requires supervision at 6 a.m. on a June morning,

you can regard it as "part of the day's work" and good experience – you will find that a career of this kind is stimulating and interesting.

(ibid.: 7)

Complete biographies of three women industrial chemists, Winifred Elizabeth Adams, Kathleen Culhane (Mrs Lathbury), and Edith Ellen Humphrey are provided in online Appendix 3.

Careers in academic chemistry

In chapter 5 we described how the chemistry department at the Royal Holloway College was originally staffed by women, as were the chemistry laboratories at both Newnham and Girton Colleges. However, very few women chemists attained high rank in a co-educational college or university, Martha Whiteley (see online Appendix 3) was one of the few exceptions. There were also a few other avenues for women in academia.

Women chemists predominated in the chemistry department of the LSMW (Rayner-Canham and Rayner-Canham, 2008: 157–63). To represent this avenue, in online Appendix 3, we have included a biographical account of Sibyl Widdows, one of the pioneer women chemists at LSMW. Women's teachers' training colleges (see chapter 5) also employed women chemistry graduates. Elfreida Cornish (see online Appendix 3) took this pathway, though like some others her life-path was not linear. Cornish changed direction in later life to become a university researcher.

A third possibility was to teach the chemistry courses at one of the several tertiary-level institutions of domestic or household science (Rayner-Canham and Rayner-Canham, 2011) and we have included the biography of Margaret Seward (Mrs McKillop) in online Appendix 3. Seward taught chemistry at the most prestigious of the Household Science Departments, that of King's College for Women, Kensington.

Careers in biochemistry

In 1933 biochemist Dorothy Jordan Lloyd wrote that the biochemical sciences required some of the most demanding training of any career:

Biochemistry at most of the universities in this country is not included in the syllabus of any degree course in science, but must be studied as a post-graduate subject for a further one or two years. The preliminary training for a career in biochemistry, therefore, usually calls for four or five years of hard work. Even after this, two years spent working on a research problem in a first-class laboratory and the attainment of a PhD are a very desirable supplementary training.

(Jordan Lloyd, 1933: 20)

Yet many young women with a chemistry background from school were attracted to, and flourished in, the field of biochemistry (Long *et al.*, 2009). Some young women joined such biochemical organizations as the Lister Institute, London (including Ida Smedley Maclean, see online Appendix 3). However, it was the Cambridge University biochemistry group under F. Gowland Hopkins that specifically encouraged young women (Rayner-Canham and Rayner-Canham, 1999: 47-49). The science historian, Mary Creese, commented:

> At the time when there were practically no women research workers in any of the other university departments at Cambridge, Hopkins gave them places in his, despite the criticism which this brought him. Even in the 1920s and 1930s, when, as a Nobel laureate with a world-wide reputation he received hundreds of applications for places in his laboratory, nearly half of the posts in his Department went to women scientists.

(Creese, 1991: 296)

As an example of a chemistry student who pursued a career in biochemistry following a research position with Hopkins, we have chosen Marjory Stephenson (see online Appendix 3).

Careers in medicine

The girls' magazines of the 1880s to 1910s extolled medicine as an excellent career option for science-educated girls. An article by Edith Huntley, MD, in an 1887–8 issue of *Atalanta* opened with: 'The object of this paper is to commend the medical profession as a newly-opened and most promising career for women to the consideration of all young educated women into whose hands it may fall' (Huntley, 1887–8: 596). She also commented that, at the time, there were only three institutions accepting women:

> ... the London School of Medicine for Women, which has been in existence twelve years, and is now thoroughly established, organized, and efficient; the Edinburgh School of Medicine for Women, which is in the second year of existence, and has not yet got through its period of financial struggle and of making the

best of difficulties; and the Dublin School of Medicine, which has lately opened its doors to female students.

(ibid.: 596)

A subsequent article by Ruth Young in *Atalanta* encouraged girls to become doctors. She pointed out that it was the 'Battle of the Eighteen Seventies' when the rights of women to study medicine were established, in particular, the founding of the Henrietta Street School of Medicine for Women in 1874 (later renamed the LSMW). She concluded her commentary with a very strong statement on the rights of women to pursue a career in medicine:

> At the present time women doctors in many cases find it difficult to obtain posts as medical officers in institutions, owing, probably to a great extent, to the prejudice which the man-mind cherishes against a woman going out of her proper sphere; which proper sphere is still, alas, looked on by too many men as the frying-pan and the dust-pan; quite forgetting ... that if all women chose domestic life as their sphere of labour, the kitchens would be overcrowded, and the very polish rubbed off the furniture from the too frequent application of the duster. When the fact is recognised that it is the duty and the right of every woman to make use of her gifts and opportunities, medical women will stand as good a chance as their brother professionals of gaining the posts they are perfectly capable of filling nobly.
>
> (Young, 1898: 696)

The Girl's Own Annual of 1907–8 contained a lengthy two-part story by the writer Nora Murrell Marris, on 'How I Became a Lady Doctor' (Marris, 1907–8). Though fictionalized, it provided a very detailed account of the training for a woman doctor at the time. Similar enthusiasm for the potential of a medical career was also expressed in the pages of *The Girl's Realm*. This magazine published two articles on medical careers for girls: one in 1900–1 by Margaret M. Traill Christie, MD, BS, DPH (Christie, 1900–1); and the other, for the next generation, in 1909–10 by Elizabeth Sloan Chesser, MB (Chesser, 1909–10). In the first article, Christie reviewed the career opportunities for a woman doctor in 1900:

> There is scientific research and medical literacy work, and posts are open as lecturers, tutors, and demonstrators in women's medical schools. Nearly half the women who qualify do so with a view to the high service of medical missions abroad. Women doctors go mostly to lands where the native women refuse or

> dislike the attendance of men as doctors ... Several Poor Law
> infirmaries, Fever Hospitals, and Lunatic Asylums have one or
> more women on their staff ... Several hospitals have women as
> house surgeons, clinical assistants, or anaesthetists. These posts
> are usually unpaid, but are sought after for the sake of experience
> and honour.
>
> (Christie, 1900–1: 168)

Twenty-five years later, the opportunities for girls in medicine were still being proclaimed, especially through training at the LSMW. Winifred Cullis, the first woman professor at the LSMW (Bond, 1956), reported that by 1923, 940 women physicians had graduated from that institution (Cullis, 1925). One such graduate was Barbara Tchaykovsky, whom we have included as an example of this life-path (see online Appendix 3).

Careers in pharmacy

In 1887 pharmacist Isabella Clarke-Keer authored an article in *Atalanta* that promoted pharmacy as a career path for girls (Clarke-Keer, 1887). In addition to the availability of positions as dispensers in Britain, Clarke-Keer proposed that there were great opportunities for women pharmacists in India:

> One great opening for women pharmacists is now in India,
> where so many of the women doctors are now settling. Women
> pharmacists should accompany them, for in many instances
> in consequence of the present want of women dispensers, the
> doctors are obliged to dispense their own medicines ... the work
> should be done by the trained and qualified Pharmacist ...
>
> (ibid.: 295)

In *The Girl's Realm Annual* of 1907–8, J.E. Walden of the Westminster College of Chemistry and Pharmacy wrote an article on pharmacy as a career for women, illustrated with two photos of groups of 'girl chemists' in the laboratory (Walden, 1907–8: 395–8). He commented that:

> The girls who take up the profession are principally doctors'
> daughters, or other relatives, those who have some means, and
> yet want something to do. A great many nurses also take it up,
> for the dispensing certificate is a valuable asset to the trained
> nurse. It is, however, generally suitable for the well-educated of
> the middle classes.
>
> (ibid.: 396)

Most of the women who became qualified pharmacists had to take positions in hospitals (Jordan, 2002). Women who attempted to practise in the retail trade found that they were not always welcome, as Emily Forster, lecturer at the Westminster College of Pharmacy, described in 1916:

> Where to settle! ... The woman pharmacist has something else to weigh besides expense: it is the question of her sex, and the fact that at present she is a pioneer in her profession, and must naturally turn to where she thinks an enterprising woman will be respected and her ability made use of rather than to a locality that appears very "Early Victorian" ... The places to avoid are centres, such as cathedral towns, where anything new is looked upon with suspicion, and must stand the test of time before it can be trusted.
>
> (Forster, 1916: 158)

As an example of a chemistry student choosing a career in pharmacy, we have included the story of Margaret Buchanan in online Appendix 3.

In this chapter we have shown that not only were girls taking chemistry because they were passionate about the joys of the chemistry lab, but also because they were encouraged to see their futures as involving careers. To have a career was a goal in itself. In addition, with the catastrophic male casualties in the First World War, they were being warned that a career was the only future that lay before most of them. With a chemistry background, they were told, they would be able to find opportunities in industrial chemistry, biochemistry, medicine, and pharmacy, in particular. But as we show in the final chapter, things would change in the late 1920s.

Chapter 12

The 1930s: The end of an era

*Today, in a good many quarters of the field, the battle we had
thought won is going badly against us – we are retreating where
once we advanced; in the eyes of certain modern statesmen women
are not personalities – they are reproductive faculty personified.*

(Hamilton, 1935: 251)

So spoke the feminist journalist Cicely Mary Hamilton. One might imagine
from the contents of the previous chapters that science, and chemistry in
particular, had become entrenched permanently in independent girls' schools
and the challenges of finding relevant careers had, over time, been solved.
This was not the case. Gillian Avery, in her history of girls' independent
schools, commented on the inter-war decline in academic focus, citing
Cheltenham Ladies' College as an example: 'Cheltenham, for instance, had
started out with high intellectual ideals, but by the 1920s these had waned'
(Avery, 1991: 12). In fact, as we noted in the Introduction, the chemistry
fervour of the 1880s–1920s passed from schools' collective memories and
it became conventional wisdom that chemistry only began in girls' schools
in the 1950s.

There are many facets as to how and why the 'retreat', as Hamilton
expressed it, came about. Though we focus on the circumstances of
chemistry for girls, it was a much broader societal (or more accurately, male)
reversal to expectations of traditional roles for women. Carol Dyhouse has
succinctly identified this difference between modern historical perception
and reality in the context of university education for women.

Most of the historians who have considered the impact of women's
admission to the universities between 1880 and 1939 have drawn
a distinction between the early pioneers, who established a right
to entry, and a later period of acceptance and integration ... any
tendency towards a narrative of steady progress makes it more
difficult to account for the conflicts of the 1920s and 1930s,
except in simple terms of backlash or stagnation.

(Dyhouse, 1995: 189)

The obvious question to be asked at this point is: why did the teaching of chemistry lose its primacy at independent girls' schools during the inter-war period? In part it seems to have been the societal factors that will be discussed later, but a key role must have been played by the Hadow Report of 1923, or to give its full name: *Report of the Consultative Committee on Differentiation of the Curriculum for Boys and Girls Respectively in Secondary Schools* (Anon., 1923).

The Hadow Report of 1923

The Taunton Commission and Bryce Commission Reports have been discussed and analysed in many contexts (see chapter 1). However, little critical study of the Hadow Report of 1923 (there were several 'Hadow Reports' of different years) could be found, even though its recommendations must have been influential, particularly in the context of girls' schools. By the very title of the report, the outcome was predetermined. Whereas in the 1880s and 1890s the goal had been to give girls the same educational opportunities as boys, by the 1920s and 1930s 'differentiation' dominated the discourse. In fact the committee was required to answer the question: 'Whether greater differentiation is desirable in the curriculum for boys and girls respectively in Secondary Schools?' (ibid.: ii).

There were 21 people on the committee, of whom only four were women. Each woman was appointed to represent a specific constituency: Essie Ruth Conway, headmistress of Tiber Street Council School, Liverpool, represented the perspectives of women elementary teachers; Emmeline Tanner, headmistress of Bedford High School for Girls, represented girls' secondary schools; Freda Hawtrey, principal of Darlington College and later of Avery Hill College, represented teachers' training colleges; and Bertha Phillpotts, principal of Westfield College, London University, represented women at university (Harrop, 2000: 164).

The commissioners saw themselves as forging a new direction for girls' education, different from that of boys. As this report is so crucial to our discourse, we consider it important to quote extensively from the document, especially this part of the introduction:

> In the first place, we feel that the education of girls and women has passed through two stages, and is, perhaps, now entering on a third. Down to 1850, and even later, it was assumed that the education of girls must be different from that of boys, because they belonged to what was regarded as the weaker (or, in a more euphemistic phrase, the gentler) sex ... During the next stage,

> which is perhaps drawing to a close, ... educational reformers claimed, and sought to secure, that there should be no difference between the education of girls and that of boys ... We may now be entering on a third stage, ... a system of differentiation under which either sex seeks to multiply at a rich interest its own peculiar talents.
>
> (Anon., 1923: xiii)

A justification for differentiation becomes apparent in the later discussions of the committee:

> Our witnesses held that the curriculum was modelled too much on the requirements of those boys and girls who were preparing for University and professional examinations and failed to provide sufficient contact with practical life. We consider that there is some substance in this criticism, especially in its bearing on girls' schools, where one of the most important aims of the training, that of fitting girls for the duties of motherhood and for work in the home, has been unduly obscured by the academic trend ...
>
> (ibid.: 58)

The committee laid great store upon the medical evidence of the time that 'proved' girls were fundamentally weaker than boys. For example, the report noted that the blood of adult males had higher haemoglobin content than that of females:

> The materially lessened amount of haemoglobin in the woman's blood after puberty is significant; haemoglobin is the agent of internal respiration, the oxygen carrier of the system; and oxygen is the great liberator of energy. It is therefore evident that the male is the better prepared for a more abundant liberation of energy with less exhaustion or fatigue.
>
> (ibid.: 82)

Following from this, it was concluded that a lesser quantity and rate of academic education was preferable for girls. Having given the 'scientific' basis for their contention of female physical inferiority, the writers of the report continued:

> It appears to be generally recognised that girls in general are not so strong physically as boys and are more highly strung and liable to nervous strain. Moreover, medical statistics seem to indicate that there is a higher percentage among girl pupils of cases of

anaemia, spinal curvature, defective eyesight, and minor physical defects. It should be added that these defects are sometimes caused and often accentuated by sedentary occupations such as needlework.

(ibid.: 85)

The commission then addressed the mental inferiority of women, including in their discourse that:

It is significant from the psychological standpoint that up to the present, despite ample opportunities, no first class genius on the creative side ... has appeared among women ... Again, in science very few women have attained to the first rank ...

(ibid.: 86)

Each subject area is discussed in the report. Under *Science*, the commission noted that a contributing factor for poor performance by girls in the physical sciences was that: ' ... the teaching is usually not so good; and there is at present a noticeable shortage of properly qualified women teachers of these subjects' (ibid.: 104). Nevertheless, the commission recommended that biology, not chemistry or physics, was the most appropriate science for girls. In their explanation, they stated:

A special reason is the comparative lack in girls of an attitude of scepticism and curiosity which gives the best approach to Natural Science. Girls have, however, an aptitude for the Biological Sciences, in which they are helped by their greater diligence and neatness; they excel in subjects which require descriptive powers and a capacity for comprehending elaborate classification.

(ibid.: 104)

The different era of the 1930s

The recommendation of the Hadow Commission Report that biology was more appropriate for girls than chemistry and physics seems to have become the 'new wisdom'. In many of the independent girls' schools in our study, the shift in emphasis from chemistry to biology during the inter-war period was apparent or could be inferred. The most comprehensive and chronological account of this change was found in the minutes of the Girls' Public Day School Company (GPDSC).

Changed emphasis at GPDSC schools

As we have shown in earlier chapters there had been a strong emphasis on the teaching of the physical sciences, particularly chemistry, at the GPDSC schools. This item in the GPDSC minutes of 1917 stated: 'It is desirable and practicable to ground every girl in Elementary Science, particularly Chemistry and Physics ...' (Anon., 1917: 20). The point was made even more strongly in 1922: 'Biology should be dropped in favour of more Chemistry and Physics, if the girls were to enter the Advanced Course as properly equipped for Science as boys' (Anon., 1922: 35).

The focus changed considerably by 1930 when there was a discussion by the GPDSC headmistresses on the place of biology in education:

> Miss Esdaile urged that Biology should be a compulsory subject, especially as such a small percentage of girls went on to Universities. Everyone should know the natural laws governing plants and animals. The [GPDSC] Trust were behind the Schools of other countries in this respect. There was no practical difficulty in keeping certain live things in Schools, such as Bees, Ants' Nests, etc. Miss Haig Brown agreed that Biology was the best subject for girls not going to a University ... Miss M.E. Lewis and Miss Cossey said that Biology developed thought along interesting lines, made girls healthy and natural, and fitted them for public health work and social life.
>
> (Anon., 1930: 136)

Not only was the emphasis shifting to biology, but also chemistry was being replaced by a more general physical science programme, as this report to the GPDSC from Nottingham Girls' High School indicated: '... the introduction of more General Science in the Middle School, the substitution in the London School Certificate of the Elementary Science paper in Chemistry and Physics for the specialized paper in Chemistry ...' (Anon., 1934: 20).

Changes in philosophy of girls' education

It was not just a shift from chemistry to biology, but it was a whole philosophical change. By the 1920s and 1930s those charismatic pioneer headmistresses, fired with the fervour of the need to match or exceed boys' schools and to provide a springboard to university, had retired or died.

To the new pragmatic generation of headmistresses, chemistry and its associated laboratory work was no longer the high priority item that it had once been. As Felicity Hunt has commented:

In the 1920s and 1930s it was fashionable to accuse girls' secondary schools of neglecting the 'feminine' side of their pupils' development. The Victorian pioneers (and Miss Buss and Miss Beale were frequently cited on these occasions) were supposed to have adopted a model of 'liberal education' and in doing so had 'assimilated' the 'boys' curriculum' and ignored the needs of femininity in their schools. The result, said the accusers, was that girls' education was a 'slavish imitation' of boys' and by definition, therefore, inappropriate for girls.

(Hunt, 1987: 3)

Even as early as 1911, Margaret A. Gilliland, headmistress of Haberdashers' Aske's Girls School, was proclaiming the new philosophy of girls' education:

In recent years there has been a widespread movement to bring the education of our girls into relation with their work as home-makers. The old "blue-stocking" type, who prided herself on not knowing how to sew or mend, and who thought cooking menial and beneath her, no longer appeals to anyone ... We want our girls to grow up into sensible, methodical, practical women, able to direct intelligently and sympathetically the manifold duties of the home ...

(Gilliland, 1911: 153)

Though at the time Gilliland's comments were contrary to those of the other headmistresses, her sentiments subsequently became more generally accepted. Dyhouse has noted:

A strong implication in the writings of these three [Margaret Gilliland, Sara Burstall, and Alice Ravenhill] and countless others of similar persuasion is that too much liking for intellectual work is selfish in women, and unacceptably so ... Sara Burstall, for instance, drew attention to the ways in which Darwinistic and 'evolutionary ideas' had demonstrated the necessity of women fulfilling their obligations to the race, judging that this must inevitably modify the goals of educators ... It was now imperative that women should learn to subordinate private interests to those of National Welfare.

(Dyhouse, 1981: 163)

The inter-war era in fiction for girls

The decline in enthusiasm for feminist causes was also evident in fiction for girls. In chapter 6 we described the pioneering role of the novels by Angela Brazil. One of her works we cited was that of *The Madcap of the School*, and of the science teacher, Miss Gibbs. Miss Gibbs was of the suffragette era, while Brazil suggested that even in 1917 her young students saw the world in a very different light, as this quote shows:

> She [Miss Gibbs] was determined ... to turn her pupils out into the world, a little band of ardent thinkers, keen-witted, self-sacrificing, logical, anxious for the development of their sex, yearning for careers, in fact the vanguard of a new womanhood ... They listened to her impassioned addresses on women's suffrage without a spark of animation, and sat stolidly while she descanted upon the bad conditions of labour among munition girls, and the need for lady welfare workers. The fact was that her pupils did not care an atom about the position of their sex, a half-holiday was more to them than the vote and their own grievances loomed larger than those of factory hands.
>
> (Brazil, 1917: 59–60)

Women's employment in the inter-war period

As Dyhouse commented above, there was no steady progress – especially when it came to employment for the young women whose education focused on the sciences. The headmistress at Howell's School, Llandaff, addressed the issue in her school report of 1926 in *Howell's School Magazine*. Interestingly, one of the new careers she identified later in the report was that of 'chemical research':

> It is increasingly difficult to find Careers for our girls. One notable difference between the present times and the days of my youth is the fact that every girl now sets forth to have a Career, whereas formerly those girls who took up Medicine or Nursing or Teaching were the exceptions; as a result, all the professions open to women are full ... There are, however, new avenues of work for the girl who has a strong and willing heart, a trained mind and a desire to follow the ancient precept "whatsoever thy hand findeth to do, do it with all thy might."
>
> (Trotter, 1926: 6–7)

Perhaps of greater importance was the atmosphere of the times. The feminist author Irene Clephane described in 1935 how public sentiment towards working women had changed dramatically in the 1920s:

> As the men of the forces were released, they resumed their places in industry and business, and the women, whose able fulfilment of their duties had won them ecstatic praise, found themselves no longer wanted in a changed world where all the doors that had so miraculously opened to them were relentlessly closing again. From being the saviours of the nation, women in employment were degraded in the public press to the position of ruthless self-seekers depriving men and their dependents of a livelihood. The woman who had no one to support her, the woman who herself had dependents, the woman who had no necessity, save that of the urge to personal independence and integrity, to earn: all of them became, in many people's minds, objects of opprobrium ...
>
> (Clephane, 1935: 200–1)

Susan Kent commented on the passing of the era of British feminism: 'The post-war backlash against feminism extended beyond the question of women's employment; a *Kinder, Küche, Kirche* ideology stressing traditional femininity and motherhood permeated British culture' (Kent, 1988: 239).

An oft-overlooked act of parliament formalized discrimination against women following the end of the First World War (Rubin, 1989). During the war there was a demand – particularly from the trade unions and the Labour Party – that a law be introduced such that after the end of the war, the rights and jobs of conscripted workers be restored to them. As a result, in 1919, the Restoration of Pre-War Practices Act was drafted and passed into law. This act actually required employers to dismiss women from the workforce once men came home and factories switched to peacetime production.

Though Gerry Rubin contended that the firing of women workers happened irrespective of the provisions of the act, he noted:

> In this endeavour, employers were no doubt aided by the technically unproblematic implementation of restoration, by the general lack of resistance to their removal by women workers themselves (apart from the protests of a number of women's organizations), and by the insistence of the trade unions that the women be removed.
>
> (ibid.: 934)

Inter-war employment for women in teaching

In the inter-war years teaching was the most common career for women science graduates (Anon., 1936). Dyhouse commented: 'Teaching, then, be it a vocation, the only realistic option or a last resort, remained the fate of the majority of women graduates in this period [pre-1939]' (Dyhouse, 1997: 221). However, an editorial of 1932 in the *Journal of Education* proclaimed that the rightful place of woman was not as a single teacher, but married and settled:

> How often do we hear it said, "We had high hopes of her – but she married." Nevertheless women are beginning to see how empty are the Pankhurstian victories. They are weary of ambition and office life in continually inferior positions; what they really need is marriage, a home, and children. And that is right, for it is their natural function.
>
> (Anon., 1932: 560–1)

There was a concern by the authorities that young women who had been given grants towards teacher training would marry upon graduation instead. Dyhouse interviewed women who had graduated between the wars, one of whom was Kathleen Uzzell who graduated in 1933. Uzzell recalled:

> When we were first at University we were called into a room where we were told we had to swear an oath to teach for five years, but it was pointed out it was a 'moral not a legal' oath ... The promise to teach for five years meant a promise not to marry as there were no married female teachers except war widows.
>
> (Dyhouse, 1997: 217)

In academia, too, the inter-war period seems to have become tougher, rather than easier for women chemists. In 1902 Martha Whiteley (see online Appendix 3) had been given a warm welcome as the Imperial College magazine, *The Phoenix*, reported: 'At the Prize Distribution, we had the very great pleasure of watching Dr Whiteley take her place with the Staff, the first lady to occupy that position at the Royal College of Science [part of Imperial College]' ('C.E.A.S', 1904–5: 12).

However, the next woman chemist hired at Imperial College, Margaret Carlton, was more marginalized. Completing her PhD in 1925, Carlton held the rank of assistant lecturer for most of her life, being promoted to lecturer only in 1946. The science historian Hannah Gay commented: 'This limited career progress was typical for women of her generation. While several of the women working at the college in this period were acknowledged as gifted scientists, they were not seen as serious candidates for professional advancement' (Gay, 2007: 155).

The succession controversy at Royal Holloway College

In chapter 11 we showed that, in the early years, women chemists found ready employment in the chemistry departments at women-only institutions: Royal Holloway College (RHC); Bedford College; King's College for Women; and the London School of Medicine for Women (LSMW). But this dominance became eroded with time. Though the main challenges seemed to come during the late 1920s and into the 1930s, there had already been 'skirmishes' in the 1910s.

One of the early defeats was at RHC. In chapter 5 we described how E. Eleanor Field was the dominant figure in the chemistry department of the RHC from 1895 until her retirement in 1913. The assumption was that, as the co-educational universities were run by men, it was only fair that RHC would be run by women. But this was not to be.

Field's up-graded replacement position of a university professorship of chemistry was advertised in the 15 March 1913 issue of *The Times* as open to men and women on equivalent terms. Sibyl Widdows, lecturer in chemistry at LSMW (see online Appendix 3) sent a letter to many of the former RHC students expressing her opposition:

> I and several other Old Students feel rather strongly that this post which has been in the hands of women ever since the opening of the College, and which has been markedly efficiently run, ought not to pass out of their hands without some strong reason. Of course we do not want any woman to be given the post, but we think that if a woman of sufficient standing and ability applies it should be offered to her. It is becoming a very serious thing for science women the way in which the science posts in women's Colleges are gradually being placed in the hands of men when there are quite good and efficient women to fill them. It means that a woman can never obtain those opportunities for research and association with other scientists which are so necessary for their work.
>
> (Widdows, 1913: 16)

Accompanying this letter was a petition, pointing out that of the nine members of the selection committee for the post, only one was a woman. Among the members of the Vigilance Committee elected by the Royal

Holloway College Association to promote the hiring of a woman professor was Margaret Seward (Mrs McKillop), lecturer in chemistry at King's College, Women's Department (see online Appendix 3).

In response to the petition, 30 RHC graduates expressed the opinion that no gender preference should be given; 207 considered that preference should be given to a woman, of whom 156 added the caveat, 'subject to equal qualifications'; while 48 insisted that the position had to be given to a woman (Anon., 1913: 14–15).

The governors of the college continued with their original plan, and George Barger was appointed as the first professor of chemistry at RHC. Barger held the position for only one year, being succeeded as head by T.S. Moore. Moore then held the post from 1914 until 1946.

Opposition to career opportunities in medicine and pharmacy

During the closing decades of the nineteenth century, medicine and pharmacy became career options for young women who had taken chemistry at school. In London the LSMW had provided a haven both as a women-only medical training school and for its women-run chemistry department. With the outbreak of war, enrolment in the male-only London medical schools plummeted and admission of women became a necessity for the survival of many of them (Dyhouse, 1998: 115–22). At Charing Cross Hospital, the historian R.J. Minney commented: 'They came swarming in. Within a few weeks the male students were mere dots amid the fluttering skirts and flowing hair in the lecture theatre' (Minney, 1967: 153).

This victory of the opening of many of the men-only medical schools to women in the 1910s proved short-lived. One by one the bar to women was reinstated in the 1920s. For example, in the case of Charing Cross Hospital, Minney noted that with the end of the war and the return of the soldiers plus new male school graduates: 'The situation had begun to adjust itself and not long afterwards women were again barred from the School. It was not until 1948 on the University's insistence, that they were readmitted' (ibid.: 154).

By the end of the 1920s, the LSMW, now named London (Royal Free Hospital) School of Medicine, was the sole institution in the capital to freely admit women to medical studies. Dyhouse has convincingly argued that the London male-only hospitals had a masculine culture with an obsession with sports. She commented: 'It is difficult to exaggerate the importance of athletics, team sports and particularly rugby in the culture of the medical schools' (Dyhouse, 1998: 125). The very presence of women was seen as compromising that male bonding through physical prowess.

Seizing the example of the medical schools, unsuccessful attempts were made to similarly reverse the inroads of women into pharmacy. For example, in 1922, the *Pharmaceutical Journal* contained a letter from 'A Pharmacist':

> It has recently been reported to the Press that in future, the London Hospital intends to restrict its students to men only. It occurs to me that this may be a good lead to the Pharmaceutical Society and also to Colleges of Pharmacy generally. While there are so many male chemists and chemists' assistants unemployed at the present time, it seems the limit of absurdity to flood the business, or profession, with a motley horde of untrained and incompetent surplus females. Is it not practicable to eliminate this undesirable element all together?
>
> ('A Pharmacist', 1922: 208)

Inter-war employment for women in industrial chemistry

Dorothy Jordan Lloyd, MA, DSc, FIC, director of the British Leather Manufacturers' Research Association (Rayner-Canham and Rayner-Canham, 2008: 323–5) was pessimistic about opportunities for women in industrial chemistry, as she wrote in *The Phoenix*, the magazine of King Edward VI High School for Girls: 'Industrial chemistry is a career in which only the exceptional woman will make much progress' (Jordan Lloyd, 1937: 14).

In a subsequent interview in the *Journal of Careers* on the prospects for women in science in 1929, Jordan Lloyd expanded upon her beliefs that only the very best should consider an industrial career. She also described the perceived shortcomings of her own gender:

> Again, there are a number of industrial chemical posts for which second or third rate ability is sufficient, and generally speaking, a man of second or third-grade ability is to be preferred to a woman of that mental calibre, because he can be used for a wider range of duties and he is not usually, though he may be, quite as inert mentally. Women are sometimes taken for routine posts because they can be offered a lower salary than men and because they sometimes show a placid contentment in routine posts and do not crave for responsibility or any duties beyond those for which they are specifically appointed.
>
> (Anon., 1929: 19)

Nine years later, the *Journal of Careers* returned to the topic of employment for women science graduates in a series of articles which exuded pessimism. The author considered one significant deterrent to hiring women was the possibility of marriage:

> This question of marriage is undeniably a deterring factor in the employment of women scientists in industry. Firm after firm, among the large number which the *Journal of Careers* has consulted, raises it as an objection. Even a woman who did brilliant work for some years, of a quality which is still remembered by men colleagues in terms of highest praise, apparently closed the door to other women in that particular firm, for it is recorded "but she left to get married and we haven't employed a woman since."
>
> (Anon., 1938: 289)

The Hadow Report of 1923 spelled out that the days of an education for girls equal to that for boys was over. Girls' education was to be considered a separate construct. No longer would the 'boys' subject' of chemistry be a requirement for science-interested girls. Girls' talent, the report proclaimed, resided more in the memorization skills of biology.

But the Hadow Report of 1923 was a creature of its times. It was part of a wider cultural phenomenon of the reversal in fortunes for women's rights and expectations. This problem was succinctly expressed by Johanna Alberti in her account of the feminist and suffragist Elizabeth Haldane: 'But without the binding force of a single unifying issue such as suffrage, feminists in the 1920s had to return to the sisyphean task of making change at a time when political life mediated against change' (Alberti, 1990: 124).

The emphasis on chemistry as a key component of equality for girls was over. The exciting times documented in previous chapters were soon forgotten, even from the institutional memory of the schools themselves. By means of this book, we have tried to bring them back to life to pay the girls and their women chemistry teachers the honour they are due.

References

Preface

Barr, B. (1984) *Histories of Girls' Schools and Related Biographical Material.* Leicester: Librarians of Institutes and Schools of Education.

Rayner-Canham, M.F. and Rayner-Canham, G.W. (2008) *Chemistry Was Their Life: Pioneer British women chemists, 1880–1949.* London: Imperial College Press.

Introduction

Avery, G. (1991) *The Best Type of Girl: A history of girls' independent schools.* London: André Deutsch.

Beddoe, D. (1983) *Discovering Women's History: A practical manual.* London: Pandora Press.

Delamont, S. (1989) *Knowledgeable Women: Structuralism and the reproduction of elites.* London: Routledge.

Haddon, C. (1977) *Great Days and Jolly Days: The story of girls' school songs.* London: Hodder and Stoughton.

Kamm, J. (1965) *Hope Deferred: Girls' education in English history.* London: Methuen.

— (1971) *Indicative Past: A hundred years of the Girls' Public Day School Trust.* London: George Allen and Unwin.

Manthorpe, C. (1993) 'Science education in the public schools for girls in the late nineteenth century'. In Walford, G. (ed.) *The Private Schooling of Girls: Past and present.* Portland, OR: Woburn Press, 56–78.

McCulloch, G. (1987) 'School science and technology in nineteenth and twentieth century England: A guide to published sources'. *Studies in Science Education,* 14, 1–32.

Parker, J.E. (2001) 'Lydia Becker's "school for science": A challenge to domesticity'. *Women's History Review,* 10 (4), 629–49.

Rayner-Canham, M.F. and Rayner-Canham, G.W. (2008) *Chemistry Was Their Life: Pioneer British women chemists, 1880–1949.* London: Imperial College Press.

Sully, J. (2010) *Howell's School, Llandaff, 1860–2010: A legacy fulfilled.* Llandaff: Howell's School.

Vlaeminke, M. (2000) *The English Higher Grade Schools: A lost opportunity.* Abingdon: Woburn Press.

Watts, R. (2003) 'Science and women in the history of education: Expanding the archive'. *History of Education,* 32 (2), 189–199.

Chapter 1

Airlie, M.F.E.O. (1962) *Thatched with Gold: The memoirs of Mabell, Countess of Airlie*. London: Hutchinson.

Anon. (1867–8) *Schools Inquiry Commission: Report Vol. 1. Chapter 6: Girls' Schools*. London: Eyre and Spottiswoode.

— (1868) *Schools Inquiry Commission: Report by Mr J. Bryce. Vol. 9. General Reports by Assistant Commissioners. Northern Counties*. London: Eyre and Spottiswoode.

— (1872) *Report of the Endowed Schools Commissioners to the Lords of the Committee of Her Majesty's Privy Council on Education*. London: Eyre and Spottiswoode.

— (1875) *Royal Commission on Scientific Instruction and the Advancement of Science. Sixth Report*. London: Eyre and Spottiswoode.

— (1880) 'Examinations, 1880'. *Cheltenham Ladies' College Magazine*, 230–2.

— (1892) 'The Educational Times: Editorial'. *The Educational Times and Journal of the College of Preceptors*, 1 January, 45, 3–26.

— (1895) *Royal Commission on Secondary Education, Vol. 3, Minutes of Evidence*. London: Eyre and Spottiswoode.

— (1900) 'The jubilee of the Frances Mary Buss schools, April 4th, 1900'. *The Magazine of the Manchester High School*, June, 62–4.

— (1901) 'Headmistress's summary of examination reports'. *Datchelor School Magazine*, 13 (1), 16.

— (1957) *The Story of the Mary Datchelor School, 1877–1957*. London: Hodder and Stoughton.

— (2008) *The Maynard School: A celebration of 350 years, 1658–2008*. Exeter: The Maynard School.

Avery, G. (1967) chapter introduction to 'Emily Shore: Self-education'. In Avery, G. (ed.) *School Remembered: An anthology*. London: Victor Gollancz, 184.

— (1991) *The Best Type of Girl: A history of girls' independent schools*. London: André Deutsch.

Baker, T.H. (2013) *Notes on St Martin's Church and Parish*. London: Forgotten Books (originally published 1906).

Binfield, C. (1981) *Belmont's Portias: Victorian nonconformists and middle-class education for girls*. London: Dr William's Trust.

Bryant, M. (1979) *The Unexpected Revolution: A study in the history of the education of women and girls in the nineteenth century*. London: Institute of Education.

Burchell, D. (1971) *Miss Buss' Second School*. London: Frances Buss Foundation.

Burstall, S.A. (1911) *The Story of the Manchester High School for Girls, 1871–1911*. Manchester: Manchester University Press.

— (1933) *Retrospect and Prospect: Sixty years of women's education*. London: Longmans, Green.

Burstall, S.A. and Douglas, M.A. (eds) (1911) *Public Schools for Girls: A series of papers on their history, aims, and schemes of study*. London: Longmans, Green.

'C.E.R.' (1908) 'A look in at Holloway College'. *Datchelor School Magazine*, 22 (3), 5.

Clarke, D. (1991) *A Daisy in the Broom: The story of a school, 1820–1958*. Tweeddale: Julia London.

Corkran, A. (1899) 'Chat with the girl of the period'. *The Girl's Realm*, 1, 216.

Crane, B.S. (1959) 'Scientific and technical subjects in the curriculum of English secondary schools at the turn of the century'. *British Journal of Educational Studies*, 8 (1), 52–64.

Croudace, C. (1898) 'A short history of Queen's College'. In Tweedie, E. (ed.) *The First College Open to Women, Queen's College, London: Memories and records of work done, 1848–1898*. London: Queen's College, 18–33.

Davies, E. (1866) *The Higher Education of Women*. London: Alexander Strahan.

Delamont, S. (1993) 'Distant dangers and forgotten standards: Pollution control strategies in the British girls' school, 1860–1920'. *Women's History Review*, 2 (2), 233–51.

— (2004) 'Burstall, Sara Annie (1859–1939)', *Oxford Dictionary of National Biography*. Oxford: Oxford University Press, 2004. Online. www.oxforddnb.com/view/article/45782 (accessed 1 January 2016).

Dixon, D. (1998/9) 'Deprived and oppressed: Victorian and Edwardian magazines for girls'. *British Library Newspaper Library News*, 25.

— (2001) 'Children's magazines and science in the nineteenth century'. *Victorian Periodicals Review*, 34 (3), 228–38.

Douglas, M.A. and Ash, C.R. (1928) *The Godolphin School, 1726–1926*. London: Longmans, Green.

'Editor, The' (1875) 'The Royal Commission on Scientific Instruction'. *The Chemical News*, 24 September, 149.

Fletcher, S. (1980) *Feminists and Bureaucrats: A study in the development of girls' education in the nineteenth century*. Cambridge: Cambridge University Press.

— (2001) *Victorian Girls: Lord Lyttelton's daughters*. London: Hambledon and London.

Forster, M. (2004) *Significant Sisters: The grassroots of active feminism, 1839–1939*. London: Random House.

Fowler, H. (1996) 'Eleanor Mildred Sidgwick, 1845–1936'. In Shils, E. and Blacker, C. (eds) *Cambridge Women: Twelve portraits*. Cambridge: Cambridge University Press, 7–28.

France, E. (1970) *Central High School for Girls, 1844–1970*. London: Central High School for Girls Magazine Committee.

Frith, G. (1985) '"The time of your life": The meaning of the school story'. In Steedman, C., Urwin, C., and Walkerdine, V. (eds) *Language, Gender and Childhood*. London: Routledge and Keegan Paul, 113–36.

Glenday, N. and Price, M. (1974) *Reluctant Revolutionaries: A century of headmistresses, 1874–1974*. London: Pitman.

Goodman, J. (1997) 'Constructing contradiction: The power and powerlessness of women in the giving and taking of evidence in the Bryce Commission Report, 1895'. *History of Education*, 26 (3), 287–306.

Goodman, J. and Harrop, S. (2002) 'Governing ladies: Women governors of middle-class girls' schools, 1870–1925'. In Goodman, J. and Harrop, S. (eds) *Women, Educational Policy Making and Administration in England: Authoritative women since 1880*. London: Routledge, 37–55.

Goodman, J.F. (2004) 'Girls' Public Day School Company (*act.* 1872–1905)', *Oxford Dictionary of National Biography*. Oxford: Oxford University Press. Online. www.oxforddnb.com/view/theme/94164 (accessed 9 December 2016).

Harwood, H. (1959) *The History of Milton Mount College*. London: Independent Press.

Jacobs, A. (2001) '"The girls have done very decidedly better than the boys": Girls and examinations, 1860–1902'. *Journal of Educational Administration and History*, 33 (2), 120–36.

Jordan, E. (1991) 'Making good wives and mothers? The transformation of middle-class girls' education in nineteenth century Britain'. *History of Education Quarterly*, 31 (4), 439–62.

Leinster-Mackay, D. (1987) 'The endowed schools legislation, 1869–1874: Some differences of interpretation'. *Educational Studies*, 13 (3), 223–38.

Löfgren, E. (1993) *Schoolmates of the Long-Ago: Motifs and archetypes in Dorita Fairlie Bruce's boarding school stories*. Stockholm: Symposium Graduale.

Lyttelton, Duke of Devonshire (1872) 'Physical science. Letter to the Endowed Schools Commission'. Appendix 5. Letter to Royal Commission on Scientific Instruction. In *Report of the Endowed Schools Commissioners to the Lords of the Committee of Her Majesty's Privy Council on Education*. London: Eyre and Spottiswoode, 66–8.

Maclure, J.S. (2006) *England and Wales 1816 to the Present Day*. Abingdon: Routledge. Vol. 2 of *Educational Documents*. 2 vols. 2006.

Marden, O.S. (1914) 'The woman of the future: Her new opportunities and her responsibilities – what will she do with them?'. *The Girl's Realm*, 16, 449–51.

McCulloch, G. (2011) 'Sensing the realities of English middle-class education: James Bryce and the Schools Inquiry Commission, 1865–1868'. *History of Education*, 40 (5), 599–613.

McDermid, J. (1989) 'Conservative feminism and female education in the eighteenth century'. *History of Education*, 18 (4), 309–22.

Mitchell, S. (1995) *The New Girl: Girls' culture in England, 1880–1915*. New York: Columbia University Press.

Moruzi, K. (2009) 'Feminine bravery: *The Girl's Realm* (1898–1915) and the Second Boer War'. *Children's Literature Association Quarterly*, 34 (3), 241–54.

Pedersen, J.S. (1975) 'Schoolmistresses and headmistresses: Elites and education in nineteenth-century England'. *Journal of British Studies*, 15 (1), 135–62.

Reynolds, K.M. (1950) 'VI: The school and its place in girls' education'. In Scrimgeour, M.A. (ed.) *North London Collegiate School 1850–1950: A hundred years of girls' education*. Oxford: Oxford University Press, 105–38.

Reynolds, K. (1990) *Girls Only? Gender and popular children's fiction in Britain, 1880–1910*. London: Harvester Wheatsheaf.

Russell, J. and Cohn, R. (2012) *Lucy Cavendish*. Google Books: Books on Demand.

Sampson, W.A. (1908) *A History of the Red Maids' School*. Privately printed.

Siddall, M. (1999) *From Schoole to School: Changing scenes, 1699–1999*. London: Devonshire House.

Tennyson, A. (1870) *The Poetical Works of Alfred Tennyson*. New York: Harper and Brothers.

Tietze, F.I. (1957) 'Tennyson at Cambridge: A poet's introduction to the sciences'. *Transactions of the Wisconsin Academy of Sciences, Arts and Letters*, 46, 221–32.

Vanes, J. (1992) *Apparelled in Red: The history of the Red Maids' School*. Bristol: The Governors of Red Maids' School.

Vardy, W.I. (1928) *King Edward VI High School for Girls Birmingham, 1883–1925*. London: Ernest Benn.

Walford, G. (ed.) (1993) *The Private Schooling of Girls: Past and present*. London: Woburn Press.

Wikipedia contributors (2015) 'Science and art department'. *Wikipedia*, 20 September 2015. Online. https://en.wikipedia.org/w/index.php?title=Science_and_Art_Department&oldid=682006744 (accessed 29 October 2015).

Zimmern, A. (1898) *The Renaissance of Girls' Education in England: A record of fifty years' progress*. London: A.D. Innes.

Chapter 2

Alic, M. (1986) *Hypatia's Heritage: A history of women in science from antiquity through the nineteenth century*. Boston: Beacon Press.

Allen, K. and MacKinnon, A. (1998) 'Allowed and expected to be educated and intelligent: The education of Quaker girls in nineteenth century England'. *History of Education*, 27 (4), 391–402.

Anon. (1892) 'School song'. *The Mount Magazine*, n. pag.

— (1895) 'Examination: subject'. *The Mount Magazine*, December, n. pag.

— (1897) 'School events'. *The Mount Magazine*, March–April, 2.

— (1898a) 'Echoes past and present' (reprinted from *Mount Echo*, March 1891). *The Mount Magazine*, June–July, 41–3.

— (1898b) 'The land of kem'. *The Mount Magazine*, January–February, 59.

— (1913) 'Alphabet'. *The Mount Magazine*, Spring, 95.

— (1915) 'School events'. *The Mount Magazine*, 9–19.

— (1920a) 'Mount alphabet'. *The Mount Magazine*, 113.

— (1920b) 'The new laboratory'. *By Kent and Skerne*, 4.

— (1928–9) 'Scientific society report'. *The Mount Magazine*, 91–7.

Armstrong, E.V. (1938) 'Jane Marcet and her *Conversations on Chemistry*'. *Journal of Chemical Education*, 15, 53–7.

Baggs, A.P., Bolton, D.K., and Croot, P.E.C. (1985) 'Stoke Newington: education'. In Baker, T.F.T. and Elrington, C.R. (eds) *Islington and Stoke Newington parishes*. London: Victoria County History, 217–23. Vol. 8 of *A History of the County of Middlesex*. 14 vols. 1909–

Bahar, S. (2001) 'Jane Marcet and the limits of public science'. *British Journal for the History of Science*, 34, 29–49.

Baynes, H. (1940) 'Memories'. *By Kent and Skerne*, 5–6.

Booth, M. (1927–8) 'Scientific society report'. *The Mount Magazine*, 97–100.

Cantor, G. (2004) 'Real disabilities? Quaker schools as "nurseries" of science'. In Wood, P. (ed.) *Science and Dissent in England, 1688–1945*. Aldershot: Ashgate, 147–61.

Cole, W.A. (1988) *Chemical Literature, 1700–1860: A bibliography with annotations, detailed descriptions, comparisons and locations*. London: Mansel.

Davies, K.M. (1981) *Polam Hall: The story of a school*. London: G. Prodhoe.

De Morgan, S.E. (1895) *Three Score Years and Ten: Reminiscences of the late Sophia Elizabeth De Morgan*, ed. De Morgan, M.A. London: Bentley.

Dreifuss, J-J. and Sigrist, N.T. (2012) 'The making of a best-seller: Alexander and Jane Marcet's *Conversations on Chemistry*'. In Lykknes, A., Opitz, D.L., and van Tiggelen, B. (eds) *For Better or for Worse? Collaborative couples in the sciences*. Basel: Springer, 19–32.

Dudley, J. (1946) *The Life of Edward Grubb, 1854–1939*. London: James Clark.

Edgeworth, M. and Edgeworth, R.L. (1798) 'Preface'. *Practical Education, Vol. 1*. London: J. Johnson.

Fyfe, A. (2008) 'Tracts, classics and brands: Science for children in the nineteenth century'. In Briggs, J., Butts, D., and Grenby, M.O. (eds) *Popular Children's Literature in Britain*. Aldershot: Ashgate, 209–28.

Golinski, J. (1999) 'Humphry Davy's sexual chemistry'. *Configurations*, 7, 15–41.

Heath, J. (1820) Letter to her mother, Hampshire Record Office, 12M58/2.

Leach, C. (2006) 'Religion and rationality: Quaker women and science education 1790–1850'. *History of Education*, 35 (1), 69–90.

Lindee, S.M. (1991) 'The American career of Jane Marcet's *Conversations on Chemistry*, 1806-1853'. *Isis*, 82, 8–23.

Marcet, J. ('Anon.') (1817) *On Simple Bodies*. 5th ed. London: Longman, Hurst, Rees, Orme, and Brown. Vol. 1 of *Conversations on Chemistry: In which the elements of that science are familiarly explained and illustrated by experiments, in two volumes*. 2 vols. 1817.

Owen, Revd R. (1894) *The Life of Richard Owen*, Vol. 1. London: John Murray. 2 vols.

Pascoe, C.E. (1879) *Schools for Girls and Colleges for Women: A handbook of female education chiefly designed for the use of persons of the upper middle class*. London: Hardwicke and Bogue.

Penney, N. (ed.) (1911) *Journal of George Fox*, Vol. 2. Cambridge: Cambridge University Press.

Rayner-Canham, M.F. and Rayner-Canham, G.W. (2009) 'British women and chemistry from the 16th to the mid-19th century'. *Bulletin for the History of Chemistry*, 34 (2), 117–23.

Roberts, E. (ed.) (1970) *Louisa: Memories of a Quaker childhood*. London: Friends Home Service Committee.

Rossotti, H. (2006) *Chemistry in the Schoolroom, 1806: Selections from Mrs Marcet's Conversations on Chemistry*. Bloomington, IN: AuthorHouse.

Sheils, S. (2007) *Among Friends: The story of the Mount School, York*. London: James and James.

Sherman, J. (1851) *Memoir of William Allen, F.R.S.* London: Charles Gilpin.

Shirren, A.J. (1951) *The Chronicles of Fleetwood House*. London: Barnes Printers.

Smith, M.F. and Waller, E.A. (1957) *The Mount School, York: 1857–1957*. The Mount School, York.

Stewart, W.A.C. (1953) *Quakers and Education: As seen in their schools in England*. London: The Epworth Press.

Sturge, H.W. and Clark, T. (1931) *The Mount School, York: 1785–1814; 1831–1931*. London: J.M. Dent.

Todd, J. (2000) *Mary Wollstonecraft: A revolutionary life.* New York: Columbia University Press.

Tolley, K. (2003) *The Science Education of American Girls: A historical perspective.* New York: Routledge Falmer.

'Vera' (1893) 'Various poems'. *The Mount Magazine*, May, n.pag.

Wakefield, P. (1798) *Reflections on the Present Condition of the Female Sex: With suggestions for its improvement.* London: Joseph Johnson.

White, A. (1931) 'Letter to her mother, 5th day, 3rd month, 1835'. Quoted in Sturge, H. W. and Clark, T. *The Mount School, York: 1785–1814; 1831–1931.* London: J.M. Dent, 43–7.

Wollstonecraft, M. (1787) *Thoughts on the Education of Daughters: With reflections on female conduct in the more important duties of life.* London: Joseph Johnson.

Wright, S. (2011) '"I have no horror of being an old-maid": Single women in the Religious Society of Friends, 1780–1860'. *Quaker Studies*, 16, 85–104.

Chapter 3

Adams, D. (1920) 'Science club'. *Our Magazine: North London Collegiate School*, July, 40–1.

Aitken, E. (1895) 'In memoriam: An address to the science club'. *Our Magazine: North London Collegiate School*, July, 24–8.

Allen, L.D. (1913) 'In praise of *The Searchlight*'. *The Searchlight: NLCS Science Club Magazine*, 31–4.

Anon. (1875) *Twenty-First Report of the Cheltenham Ladies' College.* Cheltenham: Thomas Hailing.

— (1876) *Twenty-Third Annual Report of the Cheltenham Ladies' College.* Cheltenham: Thomas Hailing.

— (1893) *Institute of Chemistry, Minutes*, 17 February 1893.

— (1902) 'The Chronicle: Changes in the staff'. *Our Magazine: North London Collegiate School*, July, 27 (81), 46.

— (1929) 'Fifth Form Chemistry Expedition'. *NLCS Magazine*, November, 82.

— (1954) 'Death'. *NLCS Magazine*, 3, 43.

— (1961) 'In memoriam: Millicent Taylor, DSc'. *Cheltenham Ladies' College Magazine*, 144.

— (2014) 'WWWord Quiz' Online. http://wwword.com/2857/quiz/best-days-of-your-life/

Baker, W. (1962) 'Millicent Taylor, 1871–1960'. *Proceedings of the Chemical Society*, 94.

Baly, E.C.C. (1943) 'John Norman Collie, 1859–1942'. *Obituary Notices of Fellows of the Royal Society*, 4 (12), 329–56.

Beale, D. (1858) *The Student's Textbook of English and General History.* London: G. Bell.

— (1880) 'History of the College'. *Cheltenham Ladies' College Magazine*, 1, 58.

— (1890) 'The History of the Cheltenham Ladies' College'. *Cheltenham Ladies' College Magazine*, 22, 27.

— (1891) 'History of the Ladies' College'. *Cheltenham Ladies' College Magazine*, 23, 11.

Blyth, E. (1894) 'Science club'. *Our Magazine: North London Collegiate School*, November, 146.

Bunting, D.E.L. (1895) 'In memoriam: An address to the science club'. *Our Magazine: North London Collegiate School*, July, 28–30.

Burt, M.E. (ed.) (1904) *Poems That Every Child Should Know*. New York: Doubleday, Page.

Buss, F. (1889) Letter to Dorothea Beale, 13 January, Archives, NLCS.

Charity Commission (1875) *Erection of Chemical Laboratory and Museum.* Archives, NLCS.

'Chief Experimenter' (1912) 'Another experiment'. *The Searchlight: NLCS Science Club Magazine*, 26.

Clarke, A.K. (1953) *A History of Cheltenham Ladies' College*. London: Faber and Faber.

Croudace, C. (1898) 'A short history of Queen's College'. In Tweedie, E. (ed.) *The First College Open to Women, Queen's College, London: Memories and records of work done, 1848–1898*. London: Queen's College, 18–33.

Davison, A. (2009) 'Literature and Education: Victorian Educators for Girls'. Online. http://unusualhistoricals.blogspot.com/2009/05/victorian-educators-for-girls.html (accessed 12 May 2009).

Dubois, M. (1932) 'Science club'. *St Martin's Past and Present*, 11, 43.

du Pré, E. (1910) 'College reminiscences'. *Cheltenham Ladies' College Magazine*, 61, 3.

'Editor' (1912a) 'Ode to ozone'. *The Searchlight: NLCS Science Club Magazine*, 29–31.

— (1912b) 'Invitations for the Science Tea'. *The Searchlight: NLCS Science Club Magazine*, 56–7.

Evans, W.G. (2014) 'Jones, Dilys Lloyd Glynne (1857–1932)'. *Oxford Dictionary of National Biography*. Oxford: Oxford University Press. Online. www.oxforddnb.com/view/article/48526 (accessed 14 July 2015).

Eyre, J.V. (1958) *Henry Edward Armstrong, 1848–1937: The doyen of British chemists and pioneer of technical education*. London: Butterworths Scientific Publications.

Gordon, S.C. (1955) 'Studies at Queen's College, Harley Street, 1848–1868'. *British Journal of Educational Studies*, 3 (2), 144–54.

Grylls, R.G. (1948) *Queen's College, 1848–1948*. London: Routledge.

Heath, G. (1890) 'The science club'. *Our Magazine: North London Collegiate School*, March, 42–3.

Hewer, E.E. (1908) 'A visit to a sulphuric acid factory'. *Our Magazine: North London Collegiate School*, February, 33 (98), 9–10.

Hill, E.M. (1909) 'The new buildings'. *Our Magazine: North London Collegiate School*, July, 34 (102), 44–9.

Hughes, M.V. (1936) *A London Girl of the Eighties*. London: Oxford University Press.

Humphrey, E.E. (1892) 'Science club'. *Our Magazine: North London Collegiate School*, November, 142.

Innes, M. (1889) 'Chronicle'. *Cheltenham Ladies' College Magazine*, 20, 262–3.

Kamm, J. (1958) *How Different from Us: A biography of Miss Buss and Miss Beale*. London: Bodley Head.

Kaye, E. (1972) *A History of Queen's College, London, 1848–1972*. London: Chatto and Windus.

'K.N.H.H.' (1954) 'Death'. *NLCS Magazine*, 3, 44.

Lawrence, M.E. (1871) 'XXXV – Look to Jesus'. *Hymns and Poems for Very Small Children*. London: The Religious Tract Society.

Maurice, F.D. (1898) 'Original objects and aims of Queen's College'. In Tweedie, E. (ed.) *The First College Open to Women, Queen's College, London: Memories and records of work done, 1848–1898*. London: Queen's College, 1–17 (originally published 1848).

Meakin, E.B. (1890) 'The science club'. *Our Magazine: North London Collegiate School*. July, 92.

Megson, B.E. (2006) 'Aitken, Edith (1861–1940)'. *Oxford Dictionary of National Biography*. Oxford: Oxford University Press, Online. www.oxforddnb.com/view/article/58463 (accessed 14 July 2015).

Meredith, P. (1927) 'The story of a carbon atom'. *Cheltenham Ladies' College Magazine*, 11 new series, 12.

Mill, C. (1987) *Norman Collie – A Life in Two Worlds: Mountain explorer and scientist, 1859–1942*. Aberdeen: Aberdeen University Press.

Moore, M. (1940) 'Ode to the bottle of phosphorous [sic]'. *Cheltenham Ladies' College Magazine*, 12.

Mullins, R. (1893) 'The North London Collegiate School for Girls'. *Sylvia's Journal*, September, 498–503.

Newth, G.S. (1894) *A Text-Book of Inorganic Chemistry*. London: Longmans, Green.

'N.F.W.' (1941) 'Edith Aitken'. *NLCS Magazine*, March, 725–6.

NLCS (1884) *Prospectus*. Archives, NLCS.

'Paedagogus Emeritus' (1907) 'An epigram'. *Journal of Education*, 39, 74–5.

Pierce, P. (2006) *Jurassic Mary: Mary Anning and the primeval monsters*. Stroud: The History Press.

Raikes, E. (1908) *Dorothea Beale of Cheltenham*. London: Archibald Constable.

Rayner-Canham, M.F. and Rayner-Canham, G.W. (2008) *Chemistry Was Their Life: Pioneer British women chemists, 1880–1949*. London: Imperial College Press.

Ridler, A.L. (1967) *Olive Willis and Downe House: An adventure in education*. London: John Murray.

Ridley, A.E. (2012) *Frances Mary Buss, and Her Work for Education*. London: Forgotten Books (originally published 1896).

Robins, E.C. (1887) *Technical School and College Building*. London: Gilbert and Rivington.

Shenstone, W.A. (1900) *The Elements of Inorganic Chemistry for Use in Schools and Colleges*. London: Edward Arnold.

Shillito, E.H. (2010) *Dorothea Beale: Principal of the Cheltenham Ladies' College, 1858–1906*. Kessinger (originally published 1920).

Silk, F. (1913) 'An introduction to chemistry'. *The Searchlight: NLCS Science Club Magazine*, 6–7.

Soar, I. (1903) 'Science club'. *Our Magazine: North London Collegiate School*, December, 103.

Spencer, S. (2006) 'The Lady Visitors at Queen's College: From the back of the class to a seat on the Council'. *Journal of Educational Administration and History*, 36, 47–56.

Steinbach, S. (2004) *Women in England, 1760–1914: A social history*. New York: Palgrave Macmillan.

'Sub-Editor' (1913) 'Three hours practical exam'. *The Searchlight: NLCS Science Club Magazine*, 16–17.

Sutton, M. (1914) 'The boracic acid flame'. *The Searchlight: NLCS Science Club Magazine*, n.pag.

Swanwick, H.M. (1935) *I Have Been Young*. London: Gollancz.

Taylor, M. (1905) 'The New Science Wing, The Ladies' College, Cheltenham'. *School World*, June, 7, 222–3.

Tree, M.B. (1898) 'Quick, thy tablets, memory!'. In Tweedie, E. (ed.) *The First College Open to Women, Queen's College, London: Memories and records of work done, 1848–1898*. London: Queen's College, 43–5.

'Two Little Oysters' (1879) 'A tale of 1879'. *Our Magazine: North London Collegiate School for Girls*, April, 11–12.

Vaughan, J. (ed.) (1961) 'Mary Watson'. In *Somerville College Register, 1879–1959*. Oxford: Oxford University Press.

Watson, N. (2000) *And Their Works Do Follow Them: The story of North London Collegiate School, 1850–2000*. London: James and James.

Wilson, M. (1891) 'Science club'. *Our Magazine: North London Collegiate School*, March, 41–2.

Chapter 4

Aitken, E. (1898) *Educational Review reprints: The teaching of science in schools as a method of induction from the concrete*. London: The Educational Review.

Anon. (1884) 'List of books sanctioned for school libraries: Chemistry and physics'. *Minutes of the Council and Committees, Reports of Examiners, &c. for 1884*, Archives, Girls' Public Day School Trust.

— (1896) *Conference on the Teaching of Science with Especial Reference to Chemistry*, 11 June, Archives, Institute of Education.

— (1900) *Conference on the Teaching of Science*. 5 March, Archives, Institute of Education.

— (1901) 'The girls' school, miscellanea'. *Central Foundation Girls' School Magazine*, 11–13.

— (1902a) *Syllabus and Examination Schedule, Based on a Course of Instruction and Laboratory Work in Elementary Chemistry and Physics; Together with Suggestions for Teachers*. Girls' Public Day School Trust Limited, July 1902, reprinted January 1912. Archives, Institute of Education.

— (1902b) 'Training in scientific method'. *School World*, September, 328.

— (1904) 'Reviews and notices: Home and Colonial Society'. *Child Life*, 15 July, 6 (23), 172.

— (1908) 'Chronicle'. *Cheltenham Ladies' College Magazine*, 58, 261.

— (1909) 'Association of Assistant Mistresses in public secondary schools'. *Journal of Education*, 31, 149–50.

Armstrong, H.E. (1903) *The Teaching of Scientific Method and Other Papers on Education*. London: Macmillan.

Bayliss, R. (1983) 'Henry Edward Armstrong and domestic science'. *Journal of Consumer Studies and Home Economics*, 7, 299–305.

Brock, W.H. (ed.) (1973) *H.E. Armstrong and the Teaching of Science, 1880–1930*. Cambridge: Cambridge University Press.

Bryant, S. (1911) 'XI. Natural science'. In Burstall, S.A. and Douglas, M.A. (eds) *Public Schools for Girls: A series of papers on their history, aims, and schemes of study*. London: Longmans, Green, 132–51.

Burstall, S.A. (1903) 'Natural science in girls' schools'. *School World*, 5, 8–10.

Burstall, S.A. and Douglas, M.A. (eds) (1911) *Public Schools for Girls: A series of papers on their history, aims, and schemes of study*. London: Longmans, Green.

Connell, L. (1971) 'Demonstrations and individual practical work in science teaching: A review of opinions'. *School Science Review*, 52, 692–701.

Davies, J.D. (1923) 'Educational movements and methods: XI – The teaching of domestic science in secondary schools'. *Journal of Education and School World*, November, 55, 725–7.

Dyhouse, C. (1977) 'Good wives and little mothers: Social anxieties and the schoolgirl curriculum, 1880–1920'. *Oxford Review of Education*, 3, 21–35.

Faithfull, L. (1912) 'Science in girls' schools'. *School World*, December, 14, 452–3.

Fisher, W.W. (1896) *Schedule of Teaching in Chemistry*. Girls' Public Day School Company, October 1896, Archives, Institute of Education.

Flintham, A.J. (1975) 'Chemistry or cookery?' *Housecraft*, 48, 105–6.

— (1977) 'The contributions of Arthur Smithells, FRS, to science education'. *History of Education*, 6 (3), 195–208.

Fortey, I.C. (1914) 'Sexless science'. *Journal of Education*, 36, 618–20.

Freund, L. (1912) 'Science in girls' schools'. *School World*, December, 14, 453–4.

Freund, I., Hutchinson, A., and Thomas, M.B. (eds) (1904) *The Experimental Basis of Chemistry: Suggestions for a series of experiments illustrative of the fundamental principles of chemistry*. Cambridge: Cambridge University Press.

Gadesden, F. and Frood, S. (1912) 'Science in girls' schools'. *School World*, December, 14, 454.

Hall, L. and Grünbaum, I. (1910) 'Correspondence: Science teaching in girls schools'. *Journal of Education*, February, 32, 105.

— (1912) *The Chemistry of Housecraft: A primer of practical domestic science*. London: Blackie.

Hartle, H.J. (1911) 'Correspondence: Housecraft in training colleges'. *Journal of Education*, 33, 849–50.

— (1912) 'Correspondence: Experimental domestic science'. *Journal of Education*, July, 34 new series, 465.

Heath, G. (1892) 'Letters to the editor: A new course of chemical instruction'. *Nature*, 46, 540–1.

Hughes, A.M. and Stern, R. (1906) *A Method of Teaching Chemistry in Schools*. Cambridge: Cambridge University Press.

'J.B.C.' (1910) 'Our photograph'. *The Gryphon: Journal of the University of Leeds*, 13 (3), 37.

Leahy, E.M. (1912) 'Science in girls' schools'. *School World*, December, 14, 455–6.

Lees, E. (1912) 'Science in girls' schools'. *School World*, December, 14, 456–7.

Maclean, I.S. (1914) 'Careers for girls: III – Household science'. *The Educational Times*, 67, 338–9.

Malim, M.C. and Escreet, H.C. (eds) (1927) *The Book of the Blackheath High School*. London: Blackheath Press.

Manthorpe, C. (1986) 'Science or domestic science? The struggle to define an appropriate science education for girls in early 20th century England'. *History of Education*, 15, 195–213.

Palmer, B. (2012) 'Ida Freund: Teacher, educator, feminist, and chemistry textbook writer.' Online. http://vipsi.org/ipsi/journals/journals/tir/2007/July/Full%20 Journal.pdf#page=51 (accessed 19 October 2016).

Phillips, G.W. (1980) *Smile, Bow and Pass On: A biography of an avant-garde headmistress, Miss Iris M. Brooks, Malvern Girls' College*. Farnborough: St Michael's Abbey Press.

Rayner-Canham, M.F. and Rayner-Canham, G.W. (2003) 'Pounding on the doors: The fight for acceptance of British women chemists'. *Bulletin for the History of Chemistry*, 28 (2), 110–19.

— (2008) *Chemistry Was Their Life: Pioneer British women chemists, 1880–1949*. London: Imperial College Press.

— (2011) 'The teaching of tertiary-level domestic chemistry in England in the early 20th century'. *Bulletin for the History of Chemistry*, 36 (1), 35–42.

Smithells, A. (1912) 'Science in girls' schools'. *School World*, December, 14, 460.

Spencer, S. (2000) 'Advice and ambition in a girls' public day school: The case of Sutton High School, 1884–1924'. *Women's History Review*, 9 (1), 75–91.

Storr, F. (1910) 'Domestic science in the secondary school'. *School World*, 12 November, 405–8.

Turner, D.M. (1927) *History of Science Teaching in England*. London: Chapman and Hall.

van Praagh, G. (1973) *H.E. Armstrong and Science Education*. London: John Murray.

Westaway, K.M. (1934) *A History of Bedford High School for Girls*. Bedford: Hockliffe.

White, J. (1913) *First Book of Experimental Science for Girls*. n. pub.

Whyte, C.G. (1901–2) 'Famous girls' schools: XII – The Mary Datchelor School'. *The Girl's Realm*, 4 (1), 268–73.

Wood, M. (1909) 'A school course of science applied to domestic life'. *School World*, 11, 140–3.

Chapter 5

Anon. (1890) *Princess Helena College Magazine*, Easter, 9.

— (1892) 'At Brondesbury'. *Maria Grey College Magazine*, July, 5–6.

— (1897) 'The college session, 1897–98'. *Maria Grey College Magazine*, November, 2.

— (1898) 'Editorial'. *Liverpool High School Chronicle*, 11, 4.

— (1902) 'The college year, 1901–1902'. *Maria Grey College Magazine*, November, 2–3.

— (1903) 'The college year, 1902–1903'. *Maria Grey College Magazine*, November, 3.

— (1912) 'Changes in staff'. *The Paulina: The magazine of St Paul's Girls' School*, 23, 2.

— (1915) 'Extraordinary general meeting, Burlington School, London, 13 November 1915'. *Minutes of the General Meetings of the Association of Women Science Teachers*, Archives, Leeds University.

— (1921) *Association of Women Science Teachers, Report for 1921*. Archives, Leeds University, 3–16.

— (1928) 'A composite history'. *Sheffield High School Magazine*, 77, 6.

— (1931–2) 'An impending departure (to Miss Ready, Senior Science Mistress 1901–1932)'. *Nottingham Girl's High School Magazine*, 6.

— (1932a) 'Eleanor Field (1893–1913)'. *College Letter, Royal Holloway College Association*, November, 57–8.

— (1932b) 'Education for women'. *Journal of Education*, 64, 560–1.

— (1932c) *Science for Girls in Post-Primary Schools: A scheme drawn up by the North-West Branch of the Association of Women Science Teachers*, unpublished, Archives, Leeds University.

— (1944) 'Dr Boyle'. *College Letter, Royal Holloway College Association*, November, 52–4.

— (1962) *A Short History with a List of Members in January 1962*. The Association of Women Science Teachers, Archives, Leeds University.

AWST (Association of Women Science Teachers). *Minutes of the General Meetings of the Association of Women Science Teachers*, Archives, Leeds University.

Ball, M.D. (1988) 'Newnham scientists'. In Phillips, A. (ed.) *A Newnham Anthology*. 2nd ed. Cambridge: Newnham College, 76–8.

Beale, D. (1900) 'Careers for girls: V – Teaching'. *The Girl's Realm*, 2, 620–3.

Bottrall, M. (1985) *Hughes Hall, 1885–1985*. Cambridge: Cambridge University Press.

Butler, K.T. and McMorran, H.I. (eds) (1948a) 'Dorothy Marshall'. *Girton College Register, 1869–1946*. Girton College, Cambridge, 638.

— (1948b) 'Margaret Tomkinson'. *Girton College Register, 1869–1946*. Girton College, Cambridge, 652–3.

'C.C.' [Coignou, C.] (1905) 'Commencement at Trinity College, Dublin'. *Magazine of the Manchester High School*, 50–1.

Cheeseman, M.K. (1928) 'Our new school'. *St Martin's Past and Present*, 7, 24–5.

Clarke, D. (1991) *A Daisy in the Broom: The story of a school, 1820–1958*. Tweeddale: Julia London.

'D.M.P.' (1954) 'In memoriam: Mary Beatrice Thomas, 1873–1954'. *Girton Review*, Michaelmas Term, 14–24.

Edwards, E. (2001) *Women in Teacher Training Colleges, 1900–1960: A culture of femininity*. London: Routledge.

Ellsworth, E.E. (1979) *Liberators of the Female Mind: The Shirreff Sisters, educational reform and the Women's Movement*. Westport: Greenwood Press.

'E.W.' [Walsh, E.] (1914) 'Ida Freund'. *Girton Review*, 9–13.

Finlay, E. (1934) *S. Swithun's School Winchester, 1884–1934*. Winchester: Warren and Son.

Fortey, I.C. (1933) 'Elizabeth Eleanor Field'. *Newnham College Roll Letter*, 52.

Gardner, A. (1921) *A Short History of Newnham College*. Cambridge: Cambridge University Press.

'H.A.B.' (1898) 'Science'. *Bedford College London Magazine*, 36, 16.

Harris, M.M. (2001) 'Margaret M. Harris (née Jamison): Chemistry'. In Crook, J.M. (ed.) *Bedford College: Memories of 150 years*. Egham: Royal Holloway and Bedford College, 81–94.

Hill, M. and Dronsfield, A. (2004) 'Ida Freund: Pioneer of women's education in chemistry'. *Education in Chemistry*, 41 (5), 136–7.

Holt, C.D. (1987a) 'Holt, C.D. to Holt, L., letter, 12 October 1889'. In Cockburn, E.O. (ed.) *Letters from Newnham College, 1889–1892: Catherine Durning Holt*. Cambridge: Newnham College, 1–3.

— (1987b) 'Holt, C.D. to Holt, L., letter, 31 October 1889'. In Cockburn, E.O. (ed.) *Letters from Newnham College, 1889–1892: Catherine Durning Holt*. Cambridge: Newnham College, 10–13.

Howarth, J. (1987) 'Science education in late-Victorian Oxford: A curious case of failure?'. *English Historical Review*, 102, 334–71.

Jenkins, E. (2009) 'The "Yorkshire Branch" of the Association of Women Science Teachers, 1926–63'. In his *75 Years and More: The Association for Science Education in Yorkshire*. Hatfield: Association for Science Education.

— (2013) 'Creating an Association for Science Education'. In Jenkins, E. and Wood-Robinson, V. (eds) *Advancing Science Education: The first fifty years of the Association for Science Education*. Hatfield: Association for Science Education.

Jewell, H.M. (1976) *A School of Unusual Excellence: Leeds Girls' High School 1876–1976*. Leeds: Leeds Girls' High School.

Kennedy, C.L. (1908) 'Edith Little'. *Leeds Girls' High School Magazine*, Winter, 2 (29), 4–5.

Kidd, W.M. (1916) 'The supply of teachers in secondary schools for girls'. *School World*, July, 18, 267.

Layton, D. (1984) *Interpreters of Science: History of the Association for Science Education*. London: John Murray.

Lilley, I.M. (1981) *Maria Grey College, 1878–1976*. Twickenham: West London Institute of Education.

Martin, G. (2011) *Hughes Hall Cambridge, 1885–2010*. London: Third Millennium and Hughes Hall.

McWilliams-Tullberg, R. (1975) *Women at Cambridge: A men's university – though of a mixed type*. London: Gollancz.

Montgomery, F. (1997) *Edge Hill University College*. Chichester: Phillimore.

Orange, B. (1903) 'Teaching as a profession for women'. In McKenna, E.M. (ed.) *Education and the Professions*. London: Chapman and Hall. Vol. 1 of *The Woman's Library*. 6 vols. 1903.

Parkes, S.M. (2004) *A Danger to the Men? A history of women in Trinity College Dublin, 1904–2004*. Dublin: Lilliput Press.

Rawson, S. (1899) 'Where London girls may study: I – Bedford College'. *The Girl's Realm Annual*, 1, 925–9.

Rayner-Canham, M.F. and Rayner-Canham, G.W. (2008) *Chemistry Was Their Life: Pioneer British women chemists, 1880–1949*. London: Imperial College Press.

Ridler, A.L. (1967) *Olive Willis and Downe House: An adventure in education*. London: John Murray.

Ridley, A.E. (1896) *Frances Mary Buss, and Her Work for Education*. London: Longmans, Green.

Rogers, A.M.A.H. (1938) *Degrees by Degrees: The story of the admission of Oxford women students to membership of the university*. Oxford: Oxford University Press.

Sayers, J.E. (1973) *The Fountain Unsealed: A history of the Notting Hill and Ealing High School*. Welwyn Garden City: Broadview Press.

Shorney, D. (1989) *Teachers in Training, 1906–1985: A history of Avery Hill College*. London: Thames Polytechnic.

Siddall, M. (1999) *From Schoole to School: Changing scenes, 1699–1999*. London: Devonshire House.

Simms, T.H. (1979) *Homerton College, 1695–1978: From dissenting academy to approved society in the University of Cambridge*. Cambridge: Trustees of Homerton College.

Stern, R. and Hughes, A.M. (1906) *A Method of Teaching Chemistry in Schools*. Cambridge: Cambridge University Press.

Sturge, H.W. and Clark, T. (1931) *The Mount School, York: 1785–1814; 1831–1931*. London: J.M. Dent.

Taylor, P.M. (1954) 'Miss M. B. Thomas, 1873–1954'. *Association of Women Science Teachers, Report for 1954*. Archives, Leeds University.

Vipont, E. (1959) *Ackworth School: From its foundations in 1779 to the introduction of co-education in 1946*. London: Lutterworth Press.

White, A.B. (1979) 'Y.G. Raymond'. *Newnham College Register 1871–1971*, vol. 1, *1871–1923*, Cambridge: Newnham College, 107.

Wikipedia contributors (2015) 'George Samuel Newth'. *Wikipedia*, 14 March 2015. Online. https://en.wikipedia.org/wiki/George_Samuel_Newth (accessed 20 October 2016).

Williams, B.H.G. (1988) *Berkhamsted School for Girls: A centenary history, 1888–1988*. Aylesbury: Hazel Watson and Viney.

Wilson, H. (1988) 'Miss Freund'. In Phillips, A. (ed.) *A Newnham Anthology*. 2nd ed. Cambridge: Newnham College, 71–2.

Chapter 6

Anon. (1882) *Minutes of the Council and Committees, Reports of Examiners, &c. for 1882*, Archives, Girls' Public Day School.

— (1884a) 'Science teaching in our public schools: IX – Girl's high schools'. *Journal of Education*, 6 new series, 352–3.

— (1884b) *Minutes of the Council and Committees, Reports of Examiners, &c. for 1884*, Archives, Girls' Public Day School.

— (1886) *Minutes of the Council and Committees, Reports of Examiners, &c. for 1886*, Archives, Girls' Public Day School.

— (1887a) 'Our visit to the North London Collegiate School'. *Milton Mount Magazine*, November, 13, 141–3.

— (1887b) *School Report*, Archives, Manchester High School for Girls.

— (1887c) 'Notes'. *Magazine of the Manchester High School*, July, 169.

— (1888) *Meeting of the Governors*, Archives, Manchester High School for Girls.

— (1890) *School Report*, Archives, Manchester High School for Girls.

— (1907) *Minutes of the Council and Committees, Reports of Examiners, &c. for 1907*, Archives, Girls' Public Day School Trust, 18 December.

— (1911) 'The story of the school'. *Winter Meeting of the [Notting Hill] Old Girls' Association*, March, unpublished, 27–31.

— (1920) 'School news, 1920'. *Ipswich High School Magazine*, 2.

— (1922) 'Trials of the "Lab"'. *White and Blue: The Alice Ottley School magazine*, 91, 1012.

— (1924) *Croydon High School 1874–1924*, 16.

— (1926) 'Head Mistress' letter'. *South Hampstead High School magazine*, November, 5–8.

— (1928) 'Opening of the Lawrence Science Building'. *Roedean School Magazine*, 30 (2), 96–7.

— (1932) 'The story of P.H.S'. *Portsmouth High School Magazine*, 5–6.

— (1956) 'Extracts from Reminiscences, 1909–1929'. In Cattley, M.H. *Perse School for Girls Cambridge 1881–1956: A short history of the school compiled to commemorate the 75th anniversary of the foundation*. Cambridge: W. Heffer.

— (1964) *A School Remembers: Sutton High School G.P.D.S.T, 1884–1964*. Merton: Croker Brothers.

— (1978) 'Frost, Louisa, 1898–1901'. In *Ipswich High School G.P.D.S.T. 1878–1978*. 7–8.

Bailes, H. (2000) *Once a Paulina: A history of St Paul's Girls' School*. London: James and James.

Bain, P. (1984) *St Swithun's: A centenary history*. Sussex: Phillimore.

Beale, D., Soulsby, L.H.M., and Dove, J.F. (eds) (1898) *Work and Play in Girls' Schools by Three Head Mistresses*. London: Longmans, Green.

Brazil, A. (1911) *The New Girl at St Chad's*. London: Blackie.

— (1914) *The School by the Sea*. London: Blackie.

— (1918) *A Patriotic Schoolgirl*. London: Blackie.

— (1925) *My Own Schooldays*. London: Blackie.

Broadway, C.M. and Buss, E.I. (1982) *The History of the School B.G.M.S.–D.A.H.S., 1882–1982*. Luton: White Crescent Press.

Carter, O. (1956) *History of Gateshead High School, 1876–1907 and Central Newcastle High School, 1895–1955*. London: Macmillan.

Cowe, K. (1905) 'Changes in the building: The new science rooms'. *Liverpool High School Chronicle*, July, 6.

De Brereton Evans, C. (1898) 'The teaching of chemistry'. In Beale, D., Soulsby, L.H.M., and Dove, J.F. (eds) *Work and Play in Girls' Schools by Three Head Mistresses*. London: Longmans, Green, 307–19.

Finlay, E. (1934) *S. Swithun's School Winchester, 1884–1934*. Winchester: Warren and Son.

Flint, L. (1989) *Wycombe Abbey School, 1896–1986: A partial history*. Privately printed.

Freeman, G. (1976) *The Schoolgirl Ethic: The life and work of Angela Brazil*. London: Allen Lane.

Godber, J. and Hutchins, I. (eds) (1982) *A Century of Challenge: Bedford High School, 1882 to 1982*. Biggleswade: Charles Elphick.

Gollin, B. and Buckley, L. (1924) 'The new chemistry laboratory.' *Belvedere School Liverpool Chronicle*, March, 33–4.

Grimshaw, F.E. (1900) 'Headmistress's summary of examiners' reports'. *Datchelor School Magazine*, 13 (1), 16.

Hahn, G. (1897) 'A visit to Holloway College'. *Our Magazine: North London Collegiate School for Girls*, November, 107–10.

Harwood, H. (1959) *The History of Milton Mount College*. London: Independent Press.

Jones H.M. (1885) 'The new buildings'. *Notting Hill High School Magazine*, 1, 2–6.

MacKenzie, A. (1930) 'Reminiscences'. *Belvedere School Liverpool Chronicle*, 45–7.

McNicol, M. (1901) 'A school journey to Cambridge'. *Magazine of the Manchester High School*, July, 48–9.

Miller, W.A. (1857) *Elements of Chemistry: Theoretical and practical. Part III: Organic chemistry*. London: Parker. Several revised editions between 1862 and 1880.

Noake, V. (1952) *History of the Alice Ottley School Worcester*. Worcester: Trinity Press.

Perkin, W.H. (1900) 'The teaching of chemistry and its development'. *School World*, 377–80.

Rayner-Canham, M.F. and Rayner-Canham, G.W. (2008) *Chemistry Was Their Life: Pioneer British women chemists, 1880–1949*. London: Imperial College Press.

Taylor, C.M. (1932) 'Miss Taylor's letter'. *Redland High School Magazine*, December, 24–5.

Tilden, W.A. (1884) *Introduction to the Study of Chemical Philosophy*. London: Longmans, Green.

Turner, G. (1906) 'School Notes'. *Central Foundation Girls' School Magazine*, 1 (11), 251.

Waller, J. (1899) 'The new science rooms'. *Roedean School Magazine*, Summer, 2, 84–5.

Whyte, C.G. (1900) 'Famous girls' schools, IV – Roedean'. *The Girl's Realm*, 2, 1060–6.

— (1900–1) 'Famous girls' schools: The Bedford High School'. *The Girl's Realm*, 3 (2), 868–74.

— (1901–2) 'Famous girls' schools: XII – The Mary Datchelor School'. *The Girl's Realm*, 4, 268–74.

Chapter 7

Adams, E. (1960) 'Miss Quartly'. *Clapham High School Old Girls' Society News Sheet*, 13, 2.

Anon. (1889) 'Report of the natural science circle'. *Wimbledon High School Magazine*, December, 12.

— (1890) 'Field club'. *White and Blue: Worcester High School Magazine*, 1, 37.

— (1892) 'The science circle'. *Wimbledon Hill High School Magazine*, 4, 4–5.

— (1893) 'Natural science club'. *Wimbledon House School News*, Christmas Term, 11.

— (1896–7) 'Science society'. *Winchester High School Chronicle*, 1, 15.

— (1903) 'School chronicle'. *Colston's Girls' School Magazine*, December, 2 (1), 5.

— (1904a) 'Field club'. *White and Blue: Worcester High School Magazine*, 43, 1064.

— (1904b) 'School news'. *White and Blue: Worcester High School Magazine*, 43, 1064.

— (1907) 'Science club.' *The Graham Street School Magazine, Francis Holland School*, Autumn, 12 (51), 12.

— (1911) 'Annual report of the Exeter High School scientific society'. *Maynard School Magazine*, 10, 9–10.

— (1916) 'The science society'. *Malvern College for Girls Magazine*, 20, 16–19.

— (1918) 'Science club notes'. *Wycombe Abbey Gazette*, 6 (8), 116.

— (1919) 'Announcements: Lectures'. *Roedean School Magazine*, Lent, 20 (2), 48.

— (1922) 'Triumphant!'. *Downe House Magazine*, 37, 17.

— (1923a) 'Science club'. *The Paulina: The magazine of St Paul's Girls' School*, 57, 4.

— (1923b) 'The science club'. *Wimbledon High School Magazine*, 32, 25.

— (1924a) *Science Club Minute Books, Downe House*, 1 June.

— (1924b) 'The Scientific Novelties Exhibition at King's College'. *The Magazine of the City of London School for Girls*, March, 29 (4), 79–81.

— (1925a) 'Science club'. *The Paulina: The magazine of St Paul's Girls' School*, 63, 6.

— (1925b) 'Science club'. *Downe House Magazine*, 46, 14.

— (1925c) 'School reports'. *Clapham High School Magazine*, 25, 54.

— (1925d) 'Science club'. *School Magazine, Haberdashers' Aske's Girls' School, Acton*, December, 9–10.

— (1926) 'The science club'. *Wimbledon High School Magazine*, 34, 28.

— (1928a) 'Science club'. *Streatham Hill High School Magazine*, 29, 16–17.

— (1928b) 'Science society'. *School Magazine, Notting Hill High School for Girls*, 43, 9.

— (1929) 'Science notes'. *Blackheath School Record*, 44, 6.

— (1930) 'Science club'. *Streatham Hill High School Magazine*, 32, 21.

— (1933) 'Science club notes'. *Wycombe Abbey Gazette*, 10 (5), n.pag.

— (1933–4) 'Visit to the steel works'. *Sheffield High School Magazine*, 12 (1), 8.

— (1935) 'Science club'. *Streatham Hill High School Magazine*, 38, 10.

— (1937) 'The Ionians'. *The Aquila: Magazine of the Bedford High School for Girls*, March, 10.

— (1938) 'The Ionians' exhibition'. *The Aquila: Magazine of the Bedford High School for Girls*, April, 8.

Baddeley, M. (1925) 'Visit to the gas works'. *Belvedere School Liverpool Chronicle*, March, 56–7.

Boothby, F. (1964) 'Lewis' lending library'. *Journal of Documentation*, 20 (4), 203–4.

Buckley, L. (1922) 'Science club'. *Belvedere School Liverpool Chronicle*, March, 75–6.

— (1923) 'Senior science club'. *Belvedere School Liverpool Chronicle*, March, 34.

Clayton, E. (1942) 'The chemistry club'. *Streatham and Clapham High School Magazine*, 46, 10.

Davis, M. (1929) 'Science club'. *Colston's Girls' School Magazine*, 2, 37.

Dubois, M. (1932) 'Science club'. *St Martin's Past and Present*, 11, 43.

Eborall, M. and Kaye, M. (1924) 'The science club'. *Wimbledon High School for Girls Magazine*, 32, 32–3.

'E.M.A.' (1933) 'Junior science club'. *Sutton High School Magazine*, 30.

'F.M.R.' (1918–19) 'Science society report'. *Winchester High School Chronicle*, 23, 16–17.

Forsyth, J. (1929) 'Science club report'. *The Phoenix: Magazine of King Edward VI High School, Birmingham*, 42, 29.

Freely, J. (2012) *The Flame of Miletus: The birth of science in ancient Greece (and how it changed the world)*. London: I.B. Tauris.

'H.K.L. M.A.C.' (1937) 'A visit to a sulphuric and hydrochloric acid works'. *Laurel Leaves: Edgbaston High School Magazine*, December, 31–2.

'J.O.E.' (1917–18) 'Science society report'. *Winchester High School Chronicle*, 22, 15.

Jupp, M. and Smith, E. (1920) 'The science society.' *IRIS: Magazine of the Walthamstow School for Girls*, Autumn, 1 (1) new series, 8.

'M.E.' (1922) 'Science club'. *Downe House Magazine*, 37, 16.

'M.P.L.' (1907) 'The chemistry club'. *Streatham Hill High School Magazine*, 9, 279.

— (1908) 'The science club'. *Streatham Hill High School Magazine*, 10, 314.

'M.T.' (1939) 'Science society'. *School Magazine, Notting Hill High School for Girls*, 54, 10.

Pipe, A. (1929) 'Expedition to the sulphuric acid works, Bramford'. *Ipswich High School Magazine*, 18–19.

Roper, I. (1910) 'The scientific society report'. *Exeter High School Magazine*, 9, 9.

Seaton, D. (1909) 'Field club report'. *The Paulina: The magazine of St Paul's Girls' School*, 15, 8.

Shears, J. (1928) 'The science club'. *Wimbledon High School Magazine,* 36, 34–5.

'S.L.' (1932) 'D.H.S.C'. *Downe House Magazine*, 67, 35–6.

Stevenson, W. (1930) 'The science exhibition, 1929'. *Wimbledon High School Magazine*, 33, 33–4.

Swain, M. (1921) 'The science club'. *School Magazine, Haberdashers' Aske's Girls' School, Acton*, Summer, 8–9.

Virgo, Miss (1921) 'Science club'. *Ipswich High School Magazine*, 16.

Woodard, R. (1932–3) 'Science society report'. *St Swithun's School Chronicle*, 37, 19–21.

Chapter 8

Anon. (1890) 'To the Editor'. *Wimbledon Hill School Magazine*, 2, 12.

— (1897) 'Not out'. *Wimbledon Hill School Magazine*, 9, 5.

— (1914) 'A ballad of chemistry'. *The Magazine of the City of London School for Girls*, July, 18 (2), 29–30.

— (1926) 'An Inspection at C.L.S.G – the inspector being William Shakespeare'. *The Magazine of the City of London School for Girls*, June, 30 (2), 42–5.

— (2011) '1911 Little Willies Book, Intro'. http://tinyurl.com/h5hyohm (accessed 23 December 2016).

Belloc, H. (1907) *Cautionary Tales for Children: Designed for the admonition of children between the ages of eight and fourteen years*. Eveleigh Nash: London.

— (1930) *New Cautionary Tales for Children: Verses*. London: Duckworth.

'Black and White and Red Seal' (1926) 'The chemists song'. *The Paulina: The magazine of St Paul's Girls' School*, 64, 12.

Brooke, R. (1971) *Rupert Brooke: A reappraisal and selection from his writings*, ed. Rogers, T. London: Routledge and Kegan Paul.

Brown, K. (1900) 'Lines to the "Lab"'. *Nottingham Girls' High School Magazine*, Spring Term, 23.

'Bump & Bump Ltd.' (1940) 'Experiments in rhyme'. *The Paulina: The magazine of St Paul's Girls' School*, 107, 14.

Carroll, L. (1889) *Sylvie and Bruno*. London: Macmillan.

'Chemistryer, A' (1903) 'Chemistry'. *Central Foundation Girls' School Magazine*, December, 1 (2), 46–7.

Cocks, K. Somers (1934) 'A cautionary tale'. *Roedean School Magazine*, Summer, 36 (1), 47–8.

Collingwood, U.R. (1934) 'Life of a carbon atom'. *Roedean School Magazine*, Summer, 36 (1), 48–9.

Culleton, C.A. (1995) 'Working class women's service newspapers and the First World War'. *Imperial War Museum Review*, November, 10, 4–12.

'E.A.R.' (1909) 'The humours of the chemistry room'. *The Magazine of the City of London School for Girls*, March, 13 (1), 10–11.

'E.I.' (1931) 'The scientific side of a country walk'. *Sheffield High School Magazine*, 86, 248–9.

Francis, D. (1938) 'The laboratory'. *IRIS: Magazine of the Walthamstow School for Girls*, December, 7.

'Globigerina' (1926) 'Ten little scientists'. *The Paulina: The magazine of St Paul's Girls' School*, 64, 11.

Graham, J.H.C. (1899) *Ruthless Rhymes for Heartless Homes*. London: Edward Arnold.

Hazel, B. (1948) 'Dedicated to the science staff'. *The Wimbledon High School Magazine*, 55, 10.

'J.B.' (1925) 'The science room ghost'. *White and Blue: The Alice Ottley School Magazine*, 202, 1264.

Kerr, O.C. (1865) *The Palace Beautiful and Other Stories*. New York: Carleton Press.

Kingsland, J. (2015) (Archivist, Downe House School), email communication with authors giving information on Madge Godfrey, 20 January.

Kipling, R. (1910) *Rewards and Fairies*. London: Macmillan.

Loveless, J. (1942) 'A warning; or the sad story of the worst science VIth Miss Iva ever had'. *The Wimbledon High School Magazine*, 5, 6.

Mackenzie, M. (1938) 'Little Daisy'. *NLCS Magazine*, July, 60 (189), 510.

'M.G.' (1911) 'The Lab'. *Downe House Magazine*, 5, 19–20.

Opie, I. and Opie, P. (1997) *The Oxford Dictionary of Nursery Rhymes*. 2nd ed. Oxford: Oxford University Press.

'P.B.' (1909) 'A New Girl's thoughts after her first lesson in practical chemistry'. *Bath High School Chronicle*, September, 16.

Rayner-Canham, M.F. and Rayner-Canham, G.W. (2011) 'British women, chemistry, and poetry: Some contextual examples from the 1870s to the 1940s'. *Journal of Chemical Education*, 88, 726–30.

Rodker, M. (1919) 'Oration on a science overall'. *Central Foundation Girls' School Magazine*, March, 3 (1), 30–1.

'Science Study, The' (1934) 'H_2O and all that'. *The Persean Magazine*, March, 11, 15–16.

Scott-White, A.H. (1883) *Chemical Analysis for Schools and Science Classes: Qualitative-inorganic*. London: Thomas Laurie.

Sella, A. (2007) 'Classic kit: Kipp's apparatus'. *Chemistry World*, 4 (11), 91.

Sellar, W.C. and Yeatman, R.J. (1930) *1066 and All That*. London: Methuen.

'Sixth at Home' (1929) 'Ten little lower sixth'. *Clapham High School Magazine*, 29, 40–1.

Symonds, E.M. (1899) 'Concerning amateur and school magazines'. *The Girl's Realm Annual*, 1, 1123–7.

Tapley, M. (1937) 'The wanderings of a nitrogen atom'. *Orme Girls' School Magazine*, July, n.pag.

'Troubadour' (1925) 'Why?'. *The Paulina: The magazine of St Paul's Girls' School*, 63, 9.

'V.' (1906) 'Extracts from the diary of a science apron'. *Oxford High School Magazine*, 80, 413–16.

'V2B.' (1926) 'A mishap in the science room'. *Oxford High School Magazine*, 141, 1058–9.

'Would-be Chemists, The' (1944–5) 'A chemical fairy story'. *Maynard School Magazine*, 24–5.

Wylie, V. (1926) 'The sad fate of the girl who meddled with the science experiment'. *Oxford High School Magazine*, 141, 1061–2.

Chapter 9

Adams, P. (1996) *Somerville for Women: An Oxford college, 1879–1993*. Oxford: Oxford University Press.

Anon. (1908) *Inspection and Examination, 1908: Pontypool County School (Girls)*, Gwent Archives, Ebbw Vale.

— (1924) 'School work'. In *City of Cardiff High School for Girls, 1895–1924*. Cardiff: City of Cardiff High School for Girls, 42–7.

— (1925) 'Editorial'. *The Howellian*, 2.

— (1927) 'The scientists society siftings'. *City of Cardiff High School for Girls Magazine*, December, 28.

— (1930) 'Headmistress's report'. *The Howellian*, November, 7.

— (1931–2) 'Science club report, 1930–1931'. *City of Cardiff High School for Girls Magazine*, 46.

— (1936) *Minutes: Annual General Meeting*, Welsh Branch, Association of Women Science Teachers. Richard Burton Archives, University of Swansea.

— (1947) *Jubilee: The history of the County School for Girls, Pontypool, 1897–1947*. Pontypool: The Griffin Press.

— (1948) *Register*, Royal Institute of Chemistry.

— (1960) *Howell's School, Llandaff, 1860–1960: A brief history*. Cardiff: Western Mail and Echo.

Carr, C. (1955) *The Spinning Wheel: City of Cardiff High School for Girls, 1895–1955*. Cardiff: Western Mail and Echo.

Child, M.D. (1939) 'Miss Florence Rich'. *Roedean School Magazine*, Michaelmas, 41, 8.

'E-d R-b' (1930–1) 'School auction'. *City of Cardiff High School for Girls Magazine*, 32.

Evans, W.G. (1990a) *Education and Female Emancipation: The Welsh experience, 1847–1914*. Cardiff: University of Wales Press.

— (1990b) 'The Welsh Intermediate and Technical Education Act, 1889: A centenary appreciation'. *History of Education*, 19 (3), 195–210.

Evans, W.G., Smith, R., and Jones, G.E. (2008) *Examining the Secondary Schools of Wales, 1896–2000*. Cardiff: University of Wales Press.

'F.G.' (1924) 'The field club'. In *City of Cardiff High School for Girls, 1895–1924*. Cardiff: City of Cardiff High School for Girls, 27–30.

Fritsch, F.E. (1939) 'Miss M.F. Rich'. *Nature*, 143, 845.

Hoggan, F.E. (1882) *Education for Girls in Wales*. London: Women's Printing Society.

'I Smelt B.A.D.E.G.G.S' (1938) 'Correspondence'. *The Dragon: Magazine of the Pontypool County School for Girls*, July, 18.

James, T.C. and Davis, C.W. (1956) 'Schools of chemistry in Great Britain and Ireland: XXVII – The University College of Wales'. *Journal of the Royal Institute of Chemistry*, 80, 568–74.

John, M. (1949) 'In Australia'. *Hywelian (Old Girls) Magazine*, January, 30–3.

Lewis, E.A. (1966) 'The Welsh branch of the Association of Women Science Teachers'. *Collegiate Faculty of Education Journal, University College of Swansea*, 28–31.

Magnus, L. (1923) *The Jubilee Book of the Girls' Public Day School Trust, 1873–1923*. Cambridge: Cambridge University Press.

'M.C.' (1924) 'The opening'. In *City of Cardiff High School for Girls, 1895–1924*. Cardiff: City of Cardiff High School for Girls, 15–19.

McCann, J.E. (1972) *Thomas Howell and the School at Llandaff*. Cowbridge: D. Brown.

McFarlane, M. and David, I. (1935) 'Science club'. *Howell's School Magazine*, 12, 34–5.

Morgan, D. (1944) 'Miss Winny'. *Howell's School Magazine*, 18, 13–14.

Rayner-Canham, M.F. and Rayner-Canham, G.W. (2008) *Chemistry Was Their Life: Pioneer British women chemists, 1880–1949*. London: Imperial College Press.

— (2009) 'Fight for rights'. *Chemistry World*, 6 (3), 56–9.

Roderick, G.W. (2001) '"A fair representation of all interests"? The Aberdare Report on intermediate and higher education in Wales'. *History of Education*, 30 (3), 233–50.

Sancho, M. (1938) 'Great office brings added responsibilities'. *The Dragon: Magazine of the Pontypool County School for Girls*, July, 14–15.

Sully, J. (2010) *Howell's School, Llandaff, 1860–2010: A legacy fulfilled*. Llandaff: Howell's School.

Trotter, E. (1930) 'Miss Trotter's letter'. *Howell's School Magazine*, 7, 6–9.

— (1944) 'Miss Winny'. *Howell's School Magazine*, 18, 9–13.

Wilks, F.E. (1961) 'Miss M.K. Turner'. *The Dragon: Magazine of the Pontypool County School for Girls*, July, 6.

Chapter 10

Anon. (n.d.) 'Inspectors report'. *Scottish Education Department*. n.pag.

— (1835) *Report of the Scottish Institution for the Education of Young Ladies; with an appendix containing separate reports, by the different teachers, of the course of instruction, and the system pursued, in their respective classes*. Edinburgh: Oliver and Boyd.

— (1839) *Report of the Scottish Institution for the Education of Young Ladies, Session 1838–39*. Edinburgh: Oliver and Boyd.

— (1880) *St Andrews School for Girls, Council Minutes*, 15 September, 1, n.pag.

— (1881) *St Andrews School for Girls, Council Minutes*, 20 January, 1, n.pag.

— (1901) 'S.L.S. science club'. *St Leonards School Gazette*, March, 4 (10), 483–4.

— (1908) *St Leonards School, Council Reports*, 4 July, 5, 99.

— (1910a) 'Opening of the new science buildings by Sir Ernest Shackleton'. *St Leonards School Gazette*, May, 7 (2), 28–9.

— (1910b) 'The new science buildings.' *St Leonards School Gazette*, May, 7 (2), 33–4.

— (1916) 'Inspectors report'. *Scottish Education Department*. 31 July, n.pag.

— (1918) 'Valete: Miss Auld'. *St Leonards School Gazette*, November, 9 (3), 40.

— (1919) 'Mary A. Dunbar, M.A. B.Sc'. *Merchant Maiden Magazine*, 12 (3), 28.

— (1920) 'Notes'. *Merchant Maiden Magazine*, 12 (3), 63.

— (1929) 'Changes in Staff'. *St George's Chronicle*, 94, 13–14.

— (1930) 'The new wing'. *St George's Chronicle*, 96, 14.

— (1938) 'St George's School, 1888–1938'. *St George's Jubilee Chronicle*, September, 6.

'A.Y.' (1930) 'Miss Georgina Kinnear: Headmistress, 1880–1890'. In *The Park School Glasgow, 1880–1930*. Glasgow: William Hodge, 15–35.

Crerar, R. (1923) 'Field Club'. *Merchant Maiden Magazine*, 16 (1), 8.

Cruickshank, M. (1967) 'The Argyll Commission Report, 1865–8: A landmark in Scottish education'. *British Journal of Educational Studies*, 15 (2), 133–47.

'E.C.' (1930) 'The school from within: Park School in the nineties'. In *The Park School Glasgow, 1880–1930*. Glasgow: William Hodge, 52–8.

'E.J.B.' (1954–5) 'Miss Thomson'. *St George's Chronicle*, 37.

'E.M.M.' (1930) 'P.S.S.A.'. In *The Park School Glasgow 1880–1930*. Glasgow: William Hodge, 75–8.

Evelyn, J. (1882) *Diaries and Correspondence* (ed. William Bray). London: George Bell.

Furlong, L. (1855) *New Movement in Education: What it is! 'Oral' education on the 'Scottish' system*. London: Thomas Hachard and William Gurner.

Grant, J.M. (1927) *St Leonard's School, 1877–1927*. Oxford: Oxford University Press.

Harte, N. (1979) *The Admission of Women to University College, London: A centenary lecture*. London: University College, London.

Lightwood, J. (1981) *The Park School, 1880–1980*. Glasgow: University of Glasgow.

'L.S.T. & N.A.V.' (1915) 'School notes'. *The Park School Chronicle*, November, 1 (12), 167.

Macaulay, J.S.A. (ed.) (1977) *St Leonards School, 1877–1977*. Glasgow: Blackie and Son.

'M.B.W.' (1930) 'The school from within: Round the old stove'. In *The Park School Glasgow, 1880–1930*. Glasgow: William Hodge, 45–52.

'M.M.' (1930) 'The Girls' School Company, Limited, Glasgow'. In *The Park School Glasgow, 1880–1930*. Glasgow: William Hodge, 9–14.

Moore, L. (2003) 'Young ladies' institutions: The development of secondary schools for girls in Scotland, 1833–c. 1870'. *History of Education*, 32 (3), 249–72.

Morgan, A. (1929) *Makers of Scottish Education*. London: Longmans, Green.

Neeley, K.A. (2001) *Mary Somerville: Science, illumination, and the female mind*. Cambridge: Cambridge University Press

Paterson, N. (1915) 'On letter writing'. *Merchant Maiden Magazine*, December, 8 (1), 33.

Pryde, D. (1893) *Pleasant Memories of a Busy Life*. Edinburgh: William Blackwood.

Purdie, T. (1885) *Report on Science Teaching at St Leonards by Prof. Thomas Purdie, Prof. of Chemistry, St Andrew's University at the Request of Miss Dove. 22 July 1885*.

Reid, D.B. (1836) *Rudiments of Chemistry; with Illustrations of Chemical Phenomena of Daily Life*. Edinburgh: William and Robert Chambers.

Roberts, A. (2010) *Crème de la crème: Girls' schools of Edinburgh*. London: Steve Savage.

Ross, A. (1925) 'A page from the diary of the field club'. *Merchant Maiden Magazine*, July, 17 (3), 78.

'Scientist' (1927) 'A disaster'. *St George's Chronicle*, 90, 14.

Shepley, N. (2008) *Women of Independent Mind: St George's School Edinburgh and the Campaign for Women's Education*. Edinburgh: St George's School.

Skinner, L. (1994) *A Family Unbroken, 1694–1994: The Mary Erskine School tercentenary history*. Edinburgh: The Mary Erskine School.

Stronach, A. (1900–1) 'Famous girls' schools: VI – The Edinburgh Ladies' College'. *The Girl's Realm*, 3 (1), 478–84.

Walker, M. (1938) 'St George's School, 1888–1938'. *St George's Jubilee Chronicle*, September, 6.

Young, Miss (1930) 'The school from within: The first fifty years'. In *The Park School Glasgow, 1880–1930*. Glasgow: William Hodge, 37–45.

Chapter 11

Anon. (1885) 'What to do with our girls'. *The Literary World*, 30 January, 111.

— (1916a) 'Jottings'. *Journal of Education*, 38 new series, 713.

— (1916b) 'Pupils leaving'. *Our Magazine: North London Collegiate School for Girls*, December, 93.

— (1916c) 'Former pupil notes'. *The Merchant Maiden*, 9 (1), 24–6.

— (1918) 'Concerning old pupils'. *Our Magazine: The Magazine of the North London Collegiate School for Girls*, March, 12–14.

Bond, M. (1956) 'Prof. Winifred Cullis, C.B.E.'. *Nature*, 178, 1266–7.

Chesser, E.S. (1909–10) 'Careers for girls: I – Medicine'. *The Girl's Realm*, 12, 329–30.

Chew, L. (1882) 'Technical education for women'. *Milton Mount Magazine*, July, 8, 33–5.

Christie, M.M.T. (1900–1) 'Careers for girls: IX – Medicine'. *The Girl's Realm*, 3 (1), 163–8.

Clarke, M.G. (1916) 'Some occupations for women'. *The Merchant Maiden*, 9 (1), 2–5.

Clarke-Keer, I. S. (1887) 'Employment for girls: Pharmacy'. *Atalanta*, October, 6, 295.

Creese, M.R.S. (1991) 'British women of the 19th and early 20th centuries who contributed to research in the chemical sciences'. *British Journal for the History of Science*, 24, 275–305.

Cullis, W. (1925) 'Women's opportunity for service'. *The Journal of Education and School World*, February, 57, 90.

Dove, J.F. (1907) 'The modern girl'. *S.L.S. [St Leonards School] Gazette*, 6 (6), 874–6.

'Editor, The' (1915) 'The War and women: The influence of the world-conflict on women's status and work'. *The Girl's Realm*, 17, 45–6.

Forster, E.L.B. (1916) 'The ideal neighbourhood for the woman pharmacist'. *Pharmaceutical Journal*, 97, 158.

— (1920) *Analytical Chemistry as a Profession for Women*. London: Charles Griffin.

Fry, H.F. (1918) 'A works laboratory'. *Laurel Leaves: Edgbaston High School Magazine*, 33–4.

Haldane, E.S. (1915) 'The vocational education of girls after the war'. *The School World*, 17 (203), 401–4.

Huntley, E. (1887–8) 'Employment for girls: Medicine'. *Atalanta*, 6, 596, 655.

Jordan, E. (2002) '"Suitable and remunerative employment": The feminisation of hospital dispensing in late-nineteenth century England'. *Social History of Medicine*, 15, 429–56.

Jordan Lloyd, D. (1933) 'Biochemistry as a career for women'. *The Journal of Careers and Monthly School Calendar*, 12 (135), 20–22.

Jowett, C. (1876) 'Female education: Its aims and ends'. *Milton Mount Magazine*, January, 66–9.

King, A. (1880) 'What our girls may do'. *The Girl's Own Paper*, 1, 462–3.

Long, V., Marland, H., and Freedman, R.B. (2009) 'Women at the dawn of British biochemistry'. *Biochemist*, 31 (4), 50–2.

Marris, N.M. (1907–8) 'How I became a lady doctor'. *The Girl's Own Annual*, 29, 701–9.

Maunsell, A. (1938) 'The analytical chemist in industry'. *The Cheltenham Ladies' College Magazine*, 33 (new series); 5–7.

Rayner-Canham, M.F. and Rayner-Canham, G.W. (1999) 'Hoppy's ladies'. *Chemistry in Britain*, January, 35, 47–9.

— (2008) *Chemistry Was Their Life: Pioneer British women chemists, 1880–1949.* London: Imperial College Press.

— (2011) 'The teaching of tertiary-level domestic chemistry in England in the early 20th century'. *Bulletin for the History of Chemistry*, 36 (1), 35–42.

Thorne, I. (1905) *Sketch of the Foundation and Development of the London School of Medicine for Women.* London: G. Sharrow.

Vanderbilt, A.T. (1884) *What to Do with Our Girls.* London: Houlston.

Walden, J.E. (1907–8) 'Girls as chemists: How a girl may take up the work of chemistry, with a view to keeping a pharmacy, or becoming a doctor's dispenser'. *The Girl's Realm Annual*, 10, 395–8.

Young, R. (1898) 'The medical profession as a calling for women'. *Atalanta*, 11, 694–6.

Chapter 12

Alberti, J. (1990) 'Inside out: Elizabeth Haldane as a women's suffrage survivor in the 1920s and 1930s'. *Women's Studies International Forum*, 13 (1/2), 117–25.

Anon. (1913) *College Letter, Royal Holloway College*, July, 14–26.

— (1917) *Minutes of the Council and Committees, Reports of Examiners, &c. for 1917*, 21 February, Archives, Girls' Public Day School Trust.

— (1922) *Minutes of the Council and Committees, Reports of Examiners, &c. for 1922*, 3 March, Archives, Girls' Public Day School Trust.

— (1923) *Report of the Consultative Committee on Differentiation of the Curriculum for Boys and Girls Respectively in Secondary Schools.* London: H.M. Stationery Office.

— (1929) 'Prospects for Women in Science'. *Journal of Careers*, 9 (89), 18–20.

— (1930) *Minutes of the Council and Committees, Reports of Examiners, &c. for 1930*, 4 July, Archives, Girls' Public Day School Trust.

— (1932) 'Education for women'. *Journal of Education*, 64, 560–1.

— (1934) *Minutes of the Council and Committees, Reports of Examiners, &c. for 1934*, Archives, Girls' Public Day School Trust.

— (1936) 'Demand for science graduates maintained: Women graduates still mainly teachers'. *Journal of Careers*, 9, 37–8.

— (1938) 'Prospects of employment for women science graduates: Part III – Industrial research laboratories'. *Journal of Careers*, 17 (185), 289–96.

Avery, G. (1991) *The Best Type of Girl: A history of girls' independent schools*. London: André Deutsch.

Brazil, A. (1917) *The Madcap of the School*. London: Blackie.

'C.E.A.S' [Speed, C.] (1904–5) 'From our lady correspondent'. *The Phoenix: Royal College of Science Magazine*, 17, 12.

Clephane, I. (1935) *Towards Sex Freedom: A history of the Women's Movement*. London: John Lane at the Bodley Head.

Dyhouse, C. (1981) *Girls Growing Up in Late Victorian and Edwardian England*. London: Routledge and Kegan Paul.

— (1995) *No Distinction of Sex? Women in British universities, 1870–1939*. London: UCL Press.

— (1997) 'Signing the pledge? Women's investment in university education and teacher training before 1939'. *History of Education*, 26, 207–23.

— (1998) 'Women students and the London medical schools, 1914–39: The anatomy of a masculine culture'. *Gender and History*, 10 (1), 110–32.

Gay, H. (2007) *The History of Imperial College London, 1907–2007*. London: Imperial College Press.

Gilliland, M.A. (1911) 'Chapter XII: Home arts'. In Burstall, S.A. and Douglas, M.A. (eds) *Public Schools for Girls: A series of papers on their history, aims, and schemes of study*. London: Longmans, Green, 153–65.

Hamilton, C. (1935) *Life Errant*. London: J.M. Dent.

Harrop, S. (2000) 'Committee women: Women on the Consultative Committee of the Board of Education, 1900–1944'. In Goodman, J. and Harrop, S. (eds) *Women, Educational Policy-Making and Administration in England: Authoritative women since 1880*. London: Routledge, 156–74.

Hunt, F. (1987) 'Divided aims: The educational implications of opposing ideologies in girls' secondary schooling, 1850–1940'. In Hunt, F. (ed.) *Lessons for Life: The schooling of girls and women, 1850–1950*. Oxford: Blackwell, 3–21.

Jordan Lloyd, D. (1937) 'Science as a career'. *The Phoenix: Magazine of the King Edward VI High School for Girls*, 58, 12–15.

Kent, S.K. (1988) 'The politics of sexual difference: World War I and the demise of British feminism'. *Journal of British Studies*, 27, 232–53.

Minney, R.J. (1967) *The Two Pillars of Charing Cross: The story of a famous hospital*. London: Cassell.

'Pharmacist, A' (1922) 'Letters to the Editor: Women in pharmacy'. *Pharmaceutical Journal*, 11 March, 108, 208.

Rayner-Canham, M.F. and Rayner-Canham, G.W. (2008) *Chemistry Was Their Life: Pioneer British women chemists, 1880–1949*. London: Imperial College Press.

Rubin, G.R. (1989) 'Law as a bargaining weapon: British labour and the Restoration of Pre-War Practices Act 1919'. *The Historical Journal*, 32 (4), 925–45.

Trotter, E. (1926) 'Miss Trotter's letter'. *Howell's School Magazine*, July, 6–7.

Widdows, S. (1913) *College Letter, Royal Holloway College*, July, 15–16.

Index

Numbers preceded by an A are for pages in the Appendices document (available online as a free download from www.ucl-ioe-press.com/books/history-of-education/a-chemical-passion/)